SHIPS FROM THE DEPTHS

ED RACHAL FOUNDATION NAUTICAL ARCHAEOLOGY SERIES

SHIPS FROM THE DEPTHS
Deepwater Archaeology

FREDRIK SØREIDE

TEXAS A&M UNIVERSITY PRESS
College Station

Copyright © 2011 by Fredrik Søreide.
Manufactured in China
All rights reserved
First edition

This paper meets the requirements of ANSI/NISO
Z39.48-1992 (Permanence of Paper).
Binding materials have been chosen for durability.

Library of Congress Cataloging-in-Publication Data

Søreide, Fredrik, 1968–
  Ships from the depths : deepwater archaeology / Fredrik Søreide.
    — 1st ed.
     p. cm. — (Ed Rachal Foundation nautical archaeology series)
  Includes bibliographical references and index.
  ISBN-13: 978-1-60344-218-3 (cloth : alk. paper)
  ISBN-10: 1-60344-218-9 (cloth : alk. paper) 1. Underwater archae-
    ology. I. Title. II. Series: Ed Rachal Foundation nautical archaeol-
    ogy series.

CC77.U5S675 2011
930.1028′04—dc22

                                      2010020841

CONTENTS

Abbreviations vii
Acknowledgments ix

PART ONE   A Survey of Deepwater Archaeology
    ONE   Introduction to Deepwater Archaeology   3
    TWO   Deepwater Archaeology: The Basic Tools   9
    THREE   History of Deepwater Archaeology   23

PART TWO   Developing a Methodology
    FOUR   Location of Deepwater Sites   101
    FIVE   Documentation of Deepwater Sites   115
    SIX   Excavation of Deepwater Sites   139
    SEVEN   Preservation Conditions in Deep Water   155
    EIGHT   Deepwater Archaeology: Law and Ethics   165

References 169
Index 177

ABBREVIATIONS

AUV	autonomous underwater vehicle
CAD	computer-aided design
CCD	charge coupled device camera
CHIRP	digital FM sub-bottom profiling system
DGPS	differential global positioning system
DP	dynamic positioning
DSM	direct survey method
GIS	geographic information system
GPS	global positioning system
HD	high definition
LBL	long baseline underwater positioning system
NTNU	Norwegian University of Science and Technology (Norges teknisk-naturvitenskapelige universitet)
OCS	Outer Continental Shelf (U.S.)
ROV	remotely operated vehicle
ROT	remotely operated tool
SBL	short baseline underwater positioning system
SHARPS	sonic high accuracy ranging and positioning system
SIT	silicon-intensified target camera
SSBL	super-short baseline underwater positioning system
UUV	untethered underwater vehicle
VR	virtual reality
WHOI	Woods Hole Oceanographic Institution

ACKNOWLEDGMENTS

This book has been a work in progress from the time I became involved in deepwater archaeology in the early 1990s. Since then the field has rapidly evolved as the technology to access deepwater sites became readily available, opening a new frontier for marine archaeologists. This book aims to show the current state of research in this exciting new field of science.

I am very grateful to the many people who have contributed research results and permissions to use previously published and unpublished illustrations. This book would not have been possible without the continued support of the Norwegian University of Science and Technology and my good friend and colleague Prof. Marek E. Jasinski. I am also particularly grateful to Gregory M. Cook and Brett A. Phaneuf from ProMare, Inc., for contributing the financial support required for finalizing and publishing this book.

PART ONE

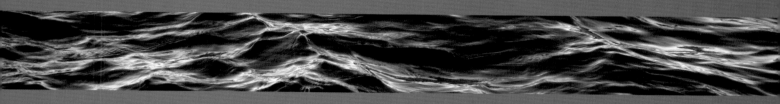

A Survey of Deepwater Archaeology

ONE

Introduction to Deepwater Archaeology

EXPLORATION OF SHIPWRECKS and archaeological sites under water started in the fifteenth century when diving suits, diving bells, and helmet diving equipment were used to salvage objects from wreck sites. The purpose of these investigations was usually solely the hunt for fine objects, though in some cases written observations and drawings, like those the Dean brothers made during their diving operations in Portsmouth harbor in the nineteenth century, were a signal of growing interest in historical content. Although such attempts made it clear that the cultural remains on the seafloor represented a large potential, archaeologists at that time were mainly forced to ignore this source material because the challenge of accessing and examining sites under water was simply too great.

Scuba equipment was first introduced in 1943, and it was only after this date that the scientific study of underwater archaeological sites began, the first of them in the late 1950s. Since then, diving archaeologists have fully developed the necessary techniques to explore and excavate underwater archaeological sites. With the advent of commercially available and inexpensive scuba systems in the past thirty years and technological advances in seafloor mapping, there has been an explosion of activity in the oceans, seas, lakes, and rivers the world over. For the first time the archaeological study of historical or ancient shipwrecks could proceed in a fashion similar to terrestrial archaeological research and excavation. Given the depth limitation of the scuba diving equipment, investigations of underwater archaeological sites have mainly been confined to shallow water (usually less than 50 m). However, since shallow water constitutes only a small percentage of the seafloor, it is likely that there are innumerable interesting archaeological sites in deep water as well, but out of reach of diving archaeologists.

Remotely operated vehicles (ROVs), originally developed more than half a century ago by the U.S. Navy to locate weapons and ships lost in depths beyond the reach of scuba divers, are now commonly used by world navies and oil and gas companies for deepwater exploration and construction. The first glimpse of these robots' enormous potential in marine archaeology came in 1989, when a team led by Robert D. Ballard and Anna Marguerite McCann used an ROV to investigate and sample a late Roman wreck more than 700 m deep near Skerki Bank, off the coast of Sicily. Since then, engineers and archaeologists have endeavored to advance from mere visual survey and random removal of artifacts with ROVs to full-blown robotic excavations. This was achieved for the first time in 2005 when a team from the Norwegian University of Science and Technology (NTNU) excavated an eighteenth-century shipwreck in 170 m depths off Norway.

Deepwater Sites

Because humankind has traveled the seas from the dawn of time, it is likely that thousands of shipwrecks and other archaeological sites lie undiscovered in deep water. These sites may contain important archaeological contributions regarding our maritime past. Along with their inherent historical value, the fact that they have been left untouched by divers and sometimes unique preservation conditions of deepwater sites makes them especially enticing for archaeologists.

Roughly seven-tenths of the earth's surface is covered by water. Continents separate this water into four main,

Helmet diver working on a wreck site (Science Museum, London)

Scuba diver working on a wreck site (NTNU Vitenskapsmuseet)

connected oceans: Arctic, Atlantic, Indian, and Pacific. The Pacific Ocean is clearly the largest, larger than all the continents combined, and also has the greatest depths (see table 1.1). These oceans hold 97 percent of the world's water; the rest is in ice and freshwater bodies. Scientists have still not explored more than a small part of the ocean's depths, since 98 percent of the seabed is beyond the reach of conventional diving. Thus, only a few deepwater sites have been located. But revolutionary developments in deep diving and robot technology now bring these deepwater sites within the reach of archaeology.

UNESCO has estimated that there are over three million shipwrecks in the world. The United Kingdom Hydrographic Office maintains a comprehensive Wrecks Database containing over 60,000 records, of which approximately 20,000 are named vessels largely in U.K. territorial waters. The NOAA's Office of Coast Survey database contains information on approximately 10,000 submerged wrecks and obstructions in the coastal waters of the United States. Norway has the longest coastline in Europe and has an estimated 20,000 shipwrecks (Søreide, 1999).

The Northern Shipwrecks Database from Northern Maritime Research contains 65,000 ship loss records for North America alone from AD 1500 to the present. The *Dictionary of Disasters at Sea during the Age of Steam* (Hocking, 1969) lists 12,542 sailing ships and war vessels lost between 1824 and 1962. And, according to the Museum of Archaeology in Lisbon, some 850 ships have gone to the bottom of the seas surrounding the Azores since 1522.

The most common reason for ships sinking is that they run aground. Statistics for the eighteenth and nineteenth centuries indicate that approximately 40 percent of all wooden sailing ships ended their careers by running onto reefs, rocks, or beaches. Contemporary statistics record that about half the sailing ships operating from the British Isles during this same period were eventually lost at sea. More than 20 percent of these sank well out, often in deep water. Statistics from Lloyds' records show

Cargo found at the "Blue China" wreck site (Odyssey)

that in the period 1864–69 almost 10 percent of all ships were lost without a trace, representing about 250 British ships every year. According to Bascom (1976), this means that approximately 10–20 percent of all seagoing ships ever built sank in deep water. Bascom also argues that, because ships were generally more seaworthy in the eighteenth and nineteenth centuries than earlier, the statistics were probably even worse for ancient ships. Based on an estimate that the Romans lost four ships to the weather for every one lost to enemy action, and knowing from written sources that warship losses numbered in the several thousand, Bascom believes that 20,000 ships from ancient times are situated in deep water in the Mediterranean alone.

Several archaeologists do not agree. Crumlin-Pedersen (1991) argues that so far only a few ancient vessels have been found along the open coast, and that they can be accounted for by special conditions. This fact is closely connected with the technique of navigation. These ships followed fixed sailing patterns, sailing coastal waters in the daylight hours and anchoring for the night in a natural harbor along the way. Wrecks are predominantly found along these sailing routes and in harbors and near market areas then in use. On occasion, ancient ships also had to leave sight of land to make an open-water crossing, and some of these ships were most likely lost during bad weather. It is, however, more likely that the majority of deep sites will relate to the age of global seafaring, when ocean crossings started for real and deepwater navigation became a necessity (Muckelroy, 1980; Throckmorton, 1991).

Since the sailing routes of these mariners changed little over the centuries, it is possible that there are areas with several shipwrecks on the seabed spanning the history of mankind. The deepwater areas that should be searched first are therefore the heavily traveled trade routes and

Table 1.1 Depths in the four major oceans
(Wilkinson, 1979)

	Average depth	Greatest depth
Arctic Ocean	990 m	4,600 m
Indian Ocean	3,890 m	7,450 m
Atlantic Ocean	3,330 m	9,144 m
Pacific Ocean	4,280 m	11,035 m

the sites of great naval battles. This is also the conclusion of Garrison (1989) for historical shipwrecks ranging from the sixteenth to twentieth century. Using statistical analyses, Garrison found that the number of wrecks increases in historical sailing routes and near major port locations and hazards such as reefs.

In addition to shipwrecks, single objects lost on purpose or accidentally from ships can be found almost everywhere on the seabed. People have also set a considerable variety of large structures in water, including fish traps, causeways, dams, docks, bridge foundations, and houses on piles or artificial islands. Eventually as these superstructures decay their remains often become totally submerged. Harbors and historical settlements have also become submerged through erosion, sea level changes, earthquakes, and volcanic activity. The most important of these sites are predominantly found in shallow water. Submerged settlements in deep water are possible in theory but have never been confirmed by archaeologists. Prehistoric sites now in deep water are especially likely in North America and North Europe because of the repeated advance and retreat of continental glaciers, which brought profound changes in sea level. When the seashore was at its lowest level, about 18,000 years ago, it was near the present continental shelf break (in North America, ca 100 m below today's level). As glaciers waned and water was released, the shoreline moved back across the continental shelf, until it reached its present level about 3,000 years ago. Several prehistoric coastal sites may therefore have vanished under water.

In northern Europe the end of the last glaciation was marked by enormous changes in the distribution of ice, sea, and land. Norwegian shelf areas were ice free for a long period before mainland Norway was, and the relative sea level may have been so low that land areas existed both in the northern North Sea and off central and north Norway. Thus, there may have existed early settlements on the shelf, more than 150 m deep today (Rokoengen and Johansen, 1996). Some stone tools have been found to support this theory, but these may just as well have been brought there with drifting ice.

Deepwater Archaeology vs. Archaeological Oceanography

The objectives of archaeology are to recover, reconstruct, and interpret the past. Archaeology is one of the oldest academic disciplines, but archaeological investigation of cultural remains under water, which is important with respect to the history of peoples, nations, and their relations with each other, is a more recent activity and still very much an area for future development.

Maritime archaeology is the archaeological subdiscipline that studies all aspects of human activity related to the sea, based on material and nonmaterial (e.g., symbolic) sources. Earlier definitions focused mainly on the surviving material evidence under water such as shipwrecks, cargo, and equipment. This is, however, the focus of **marine archaeology**, which studies human use of the sea and its resources and is particularly relevant for cultural resource management. **Nautical archaeology** is the study of ancient and historical ships.

Underwater archaeology is the archaeological field method used to investigate cultural remains under water. The terms **nautical**, **wetland**, and **waterfront archaeology** have also been used to classify different aspects of maritime archaeology, but these terms are more specific than the anthropological aspects of maritime archaeology.

Cultural resource management is the development and maintenance of programs designed to protect, preserve, scientifically study, and otherwise manage cultural resources, including prehistoric, historical, and recent sites. Cultural resources are valuable but finite, nonrenewable assets, and a nation's cultural resource base should be properly managed to achieve maximum benefit for the country's people. Underwater cultural heritage or remains means all underwater traces of human existence—buildings and other structures, artifacts, human remains, shipwrecks and their cargo or other contents, together with their archaeological and natural context.

What, then, is deepwater archaeology? Although several definitions can be made, **deepwater archaeology** is here used to denote archaeological investigations in depths that are inaccessible by scuba diving, that is, deeper than 50 m. This deeper investigation usually requires technology such as ROVs and remote sensing. Deepwater archaeology is therefore a combination of science and archaeology, which has led to the inclination to use the term **archaeological oceanography** (Ballard et al., 2007), especially in the United States.

Oceanography is by definition the study of the ocean and has four principle subdisciplines: biological, chemical, geological, and physical. Following this definition, archaeological oceanography is the interdisciplinary science aimed at locating and studying archaeological remains in the ocean. It recognizes that an interdisciplinary approach is needed, involving archaeologists, marine geologists, sedimentary bio/geochemists, engineers, and a diverse host of other marine professionals.

That being said, this definition focuses mainly on the scientific and engineering endeavors applied to archaeology. In reality there are no oceanographic archaeologists and there is no deepwater archaeology. There is archaeology that occurs in deep water indeed, but it is simply archaeology. The depth is an impediment to be overcome, requiring new technologies and methods but not "new archaeologists." To say that there is such a thing as a "deep-submergence archaeologist" is tantamount to saying that there is such a thing as a "high-altitude archaeologist"—one who can work on a site from any culture of any date, so long as it is above the tree line. It simply doesn't make sense. A site on the seafloor requires the choice of an archaeologist who has an interest in and a proficiency with the sites and archaeology of the associated culture. All in all, I prefer the definition of deepwater archaeology as archaeology that occurs in the deep ocean.

Research vs. Treasure Hunting

For many, the word **shipwreck** evokes images of pirates and buried treasure on deserted islands in the tropics. Unfortunately, that perception has carried over to underwater archaeologists. Indeed, in the early days of underwater archaeological research it was difficult for the public to discern between underwater archaeological research and treasure hunting; for some in the archaeological community itself, the dividing line may have seemed awfully thin at times as well. Initially, archaeologists were hard pressed to raise not only capital for underwater archaeological research but sufficient interest within the academic framework. Treasure hunters—for lack of a better term—

Ships in Trondheim harbor (NTNU Vitenskapsmuseet)

had things a bit easier; there were virtually no regulations protecting underwater sites, and there was no developed methodology to check their actions. The promise of striking it rich by locating a treasure ship with gold waiting on the seafloor lured investor after investor to fund often ill-fated missions. More often than not, the only financially successful individual involved with these treasure hunts was the self-styled explorer or project leader.

The unfortunate outcome was that, while underwater archaeology was maturing into an accepted subdiscipline of archaeology, valuable information about the past was lost as shipwrecks were in some cases literally torn apart in the quest for gold. That is the crux of the matter and the center of a debate raging today. Who has the right to assign the value of the information contained in a shipwreck site or the actual monetary worth of the objects therein? Who has the right to explore, salvage, or excavate? This is the basis of the dividing line between treasure hunters and marine archaeologists.

A shipwreck is essentially a snapshot of the past, capturing a moment in history without contamination from subsequent generations of inhabitants mingling with the picture, as in terrestrial sites. A shipwreck is essentially an unread book. Its value lies not in its antiquity but in the information it contains. This does not mean that all shipwrecks should be excavated and the information extracted. In fact, an equally fierce debate is raging today as to how best to preserve our heritage. Excavation is the complete and systematic destruction of an archaeological resource; all that remains is the information—with any luck gathered by a talented, well-educated crew and published. In many instances it is best to shepherd these cultural resources responsibly until such time as their excavation becomes necessary. Standards must be devised and adhered to and the information protected.

As the desire to excavate shipwrecks grew and became more feasible, the field of underwater archaeology blossomed. Academic programs offering advanced degrees appeared; nonprofit research organizations formed; maritime museums were effused with new life; and governmental agencies devised laws, regulations, and strategies to protect and investigate archaeological resources. The past thirty years have been an exciting time to be involved with underwater archaeological research, to say the least. Shipwrecks have revealed hitherto undreamed secrets about the past, and in many cases they have rewritten the history of a whole region of the ancient world.

Today the same battle is going on in deeper waters. Backed by investors, salvage companies are using advanced technology to salvage artifacts and gold from shipwrecks. In the public domain these companies claim to do archaeology. Archaeologists with no funding can only shout out their frustration. In recent decades, however, a few institutions in the United States, France, and Norway have also been able to do archaeology in the deep sea, offering an alternative to the treasure hunters. As these examples are becoming better known and the equipment and methods needed are becoming more available, archaeologists will fight back and reclaim the deep sea cultural heritage for humanity.

TWO

Deepwater Archaeology: The Basic Tools

Exploration of deepwater archaeological sites represents a meeting between the future and the past. The investigation of deepwater sites is possible only through the deployment of state-of-the-art technology.

Archaeology is concerned with the identification and interpretation of physical evidence left by past ways of life. The general activities are planning, searching for sites, documentation of sites, excavation of sites, postprocessing, and dissemination of information. The process of an underwater archaeological investigation is shaped by basic principles. These are the same regardless of water depth, but the techniques, methods, and equipment used can differ substantially, and it is much more difficult to access archaeological material in deep water.

The Process of Deepwater Archaeology

Archaeological fieldwork can be seen as the execution of several specific tasks related to discovery and investigation of sites (Dean et al., 1995; Bowens, 2008). Marine archaeological work typically consists of the five phases (Muckelroy, 1978): planning, searching for sites, documenting sites, excavating sites, and postprocessing of data.

Collection of background information (archive studies) can help focus the fieldwork activities. It is also a good idea to have a clearly defined plan, including a research and fieldwork strategy.

Since the positions of only a few deepwater sites are known, it is usually necessary to search actively for sites. There is an obvious difference between a search for a specific site and a more general survey to gain information on the nature and amount of archaeological material present in a specific area.

The documentation phase includes the construction of an accurate and complete record of the site. Collection of a few, select items to determine the age of the site is also common. Tasks include establishing the extent of the site, documenting, measuring and positioning visible objects and distances between objects, and trying to establish the extent of the buried material. The result can be a map that defines the limits of the overall site, often referred to as a predisturbance plot chart. An accurate three-dimensional picture of the site is preferable, but in practice site records are usually two-dimensional charts with supporting descriptions and measurements.

The recording process must be as objective as possible. Measurements, internal positioning of artifacts, and the relationships between objects are crucial in answering key questions about the site. Several galley bricks may, for instance, suggest a galley oven. Large numbers of small artifacts can make it necessary to record only the larger objects or groups of artifacts. Environmental evidence must also be recorded and used to interpret the site formation processes and for conservation purposes.

In shallow water, positioning and measurements of site features and artifacts are typically done by divers using tape measurements. Alternative methods are needed in deep water. The documentation and recording phase of a well-preserved and complicated site is more difficult than usual, but also more important. In some projects the information required to answer specific questions may be obtained in the documentation phase, in which case the documentation work can be the end product rather than simply a phase before excavation.

Excavation should be conducted only when other methods are inadequate or when specific circumstances require it. An excavation destroys the protection and stability of the artifacts in the seabed and is extremely expensive. Artifacts recovered during a three-month excavation can require three to five years to conserve and study. When an excavation is justified, various strategies can be employed, including test pits, trenches, or sections, which are more common and less destructive and costly than a full excavation. The method should be evaluated in terms of the balance between information retrieval and impact on the surviving remains. Excavation is not simply about gathering artifacts. During the excavation phase everything must be recorded and documented and the data interpreted with respect to both content and context—that is, what can be found, and what relationships can be discovered between the various pieces.

The post-fieldwork phase includes the conservation of artifacts, further recording, interpretation of collected data, and eventual dissemination of the results through publications, video documentation, websites, and the like.

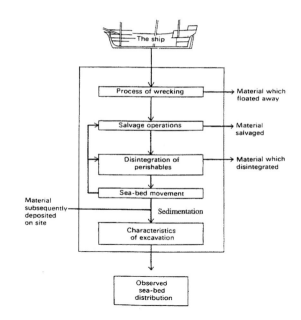

Formation processes (Muckelroy, 1978)

Site Classification

An underwater archaeological site is often a concentration of evidence in one specific place, but there is a great range and diversity. The various site formation processes include physical, biological, chemical, and man-made interference that have transformed a site to its present state.

Muckelroy (1978) has written extensively on the expectations for shipwreck distributions and preservation of specific elements of these sites. His fundamental taxonomy divides shipwrecks into continuous and discontinuous sites. The continuous sites represent shipwrecks which, while undergoing the varying kinds of deterioration mentioned above, are still relatively localized in the remains of the hull. The artifact distribution associated with these wrecks has not been much disturbed, and the sites are self-contained in a relatively small area. If the hull is missing, certain artifacts can be taken as indicators of bow and stern, and the layout of the ship may still be deduced if the site is continuous.

Discontinuous sites are those with elements of the ship widely scattered, with no single locus of the wreck site. These sites have been severely disturbed by the wrecking process. There is typically a total absence of any defining ship structure, making the reconstruction of these sites extremely difficult. The methods used to investigate discontinuous sites are obviously different from the methods used to investigate continuous sites. It is clearly much more difficult to interpret the remains of a scattered wreck site. A continuous site, on the other hand, is more difficult to investigate because of the many details. There are several intermediate site types at which the remains are neither perfectly preserved nor smashed to pieces. None of these categories vary directly with site depth.

Technology for Deepwater Archaeology

In shallow water, archaeological work is usually carried out by divers. The use of diving archaeologists is, however, much more difficult in deep water. Divers may be archaeologists who have been trained to dive, or divers who have received training in scientific techniques, that is, archaeological methods. The level of training required varies from country to country, but all deepwater diving must follow well-established codes of practice and safety procedures.

Deep diving on air to more than 30 m can lead to dive narcosis, shorter bottom times, and increased risk of decompression sickness (Arnoux, 1996). The latter problem occurs when the nitrogen that has dissolved in the blood under pressure at depth comes out of solution during ascent, creating bubbles that block veins and arteries. Several other long-term effects have also been reported,

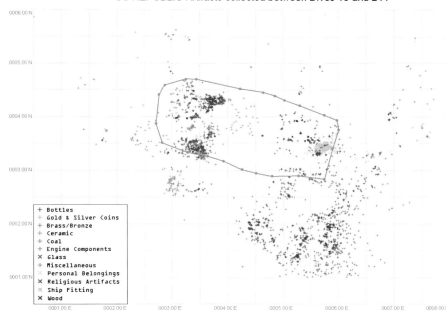

Distribution of artifacts on the SS *Republic* site (Odyssey Marine Exploration)

Deepwater diving system (Jonathan Adams)

including bone necrosis, anxiety, and possible neurological effects, but the time span of professional diving has been too short to allow construction of any firm conclusions. Other breathing gases and techniques must therefore be used when diving in deep water.

The most common deep-diving technique today is referred to as mixed-gas diving (Warrinder, 1995), in which divers breathe a mixture of gas other than naturally occurring air. Gas mixtures in use today are nitrox, a mixture of nitrogen and oxygen; heliox, a mixture of helium and oxygen; and trimix, a mixture of helium, oxygen, and nitrogen. In deep diving, it is also common to use a diving bell. The so-called bounce diving technique allows divers to be transported to the site in the bell. On site the divers are let out to complete the work, and after a set period they reenter the bell and decompression commences. When the work cannot be completed in a short period, so-called saturation diving can be used; divers are kept under pressure in chambers for the whole period, and decompression is done only at the end of the work period. Because of the complex equipment, the use of saturation diving is expensive (Allwood, 1990).

There are many examples of deepwater archaeological sites investigated by divers. In 1994 an Anglo-Swedish team began a survey-oriented nonintrusive archaeological investigation of a sixteenth-century caravel shipwrecked in the Stockholm archipelago (Adams and Rönnby, 1996). Because the site was situated in depths of 30–56 m, safety measures and the low efficiency of divers using air at these depths meant that other diving techniques had to be used (Adams, 1996). Rebreathers, which have been used by Florida State University for deepwater investigations, were considered, but because there were not enough units available for the scheduled fieldwork a system was instead developed to combine the convenience and mobility of scuba diving with the added security of surface-supplied dive equipment and a breathing mixture best suited to the working depth. Using the bounce diving technique, the divers worked from a diving bell, which functioned as a lift to and from the site as well as a decompression chamber. From 30 to 36 m depth, the divers used conventional scuba equipment and breathed enriched air. Below 36 m, the divers were breathing heliox and wearing hot-water suits. If the work had been deeper, saturation diving would have been necessary—and prohibitively more expensive.

To help solve the problems of deep diving, scientific and technical specialists have undertaken new lines of research and developed alternative equipment. This equipment has usually been developed for oceanographic, military, and industrial purposes but can also be used for marine archaeology. For example, remotely operated vehicles can be used to locate, document, and excavate archaeological sites in deep water. Compared to divers, these remote intervention systems are tools whose depth limitation is primarily a function of cost.

It is a commonly held belief that such technological innovations and refinements have greatly increased the archaeologist's capacity to do archaeology under water. However, even though existing equipment can be used to solve tasks related to marine archaeology, there are obvious flaws associated with its use. It is therefore important to use the correct equipment and develop suitable methods for its use in archaeology.

Remotely Operated Vehicles

The first ROV, *Poodle*, was developed in the United States in 1953 by Dimitri Rebikoff for the purpose of locating and filming shipwrecks (Morgan, 1990). It drastically increased the flexibility of deepwater work.

An ROV eliminates danger by keeping the operators on the surface at a lower cost and with the same capability as manned submersibles. ROVs also have the advantage that they can work in the hazardous deep-sea environment around the clock. ROVs have therefore emerged as the most flexible underwater vehicle. In general, three classes of ROVs can be identified:

1. Tethered vehicles, free swimming:
    (a) pure observation vehicles
    (b) observation vehicles with payload option
    (c) work-class vehicles
2. Bottom-crawling vehicles
3. Untethered, autonomous vehicles (AUVs)

Typical ROV system configuration
(Sperre AS)

Both the availability and capability of ROVs have been drastically improved since the 1950–60s, and today they can perform extremely advanced tasks (Whitcomb, 2000). ROVs can be seen as platforms that carry instruments to a site to document and excavate it. Archaeological tasks performed via ROV usually include general site assessment, photo and video documentation, and positioning and measurement of site features and selected objects. To complete these tasks, the ROV can be equipped with video and photo cameras, lights, sonar systems, and manipulator arms that enable it to do the work. In a few cases an ROV equipped with tools to remove sediment has also been used to excavate a site.

ROVs are a continuum of types and capabilities, from simple video carriers to multipurpose vehicles with any number of options (Bell et al., 1994; Gallimore and Madsen, 1996). ROVs can be custom-made for a particular application, but it is common to use a standard vehicle system and arm it with specialized ROV tools for the tasks in question. Most archaeological investigations have used standard industry or research ROVs equipped with special tools. The diver's senses, vision and touch, have been replaced by an electronic display (Hummel, 1995).

More than a thousand robotic, uninhabited undersea vehicles are presently in regular operation worldwide. Most are commercial ROVs designed to perform subsea inspection, survey, construction, and repair operations at modest (less than 1,000 m) depths. The basic components of all ROV systems—power supply, control console, tether, and the vehicle itself—vary in their physical dimensions. A typical vehicle consists of a frame; a propulsion system that gives the ROV maneuverability; a control system; and a buoyancy module to obtain the desired buoyancy, which is usually close to neutral. The tether transmits power to the vehicle and communication in each direction. In addition, most systems have additional control units for various equipment and a tether management system consisting of a winch and a handling system.

Remotely operated systems can be equipped with various remote sensing equipment such as sidescan sonar, magnetometers, and sub-bottom profilers to locate and document sites (Bell and Nowak, 1993; Gearhart, 2004).

Observation-class ROVs

Most low-cost observation ROVs have important limitations. The design of small observation ROVs must include a trade-off between weight/volume, cost, and working capability. They are small and not very powerful and therefore have a limited ability to move in strong currents. Having limited power also limits the practical working depth. Unless the system is operated with a tether management system, the limited power becomes insufficient as the drag on the tether increases with water depth. Small low-cost ROV systems are typically also relatively unstable due to the tether drag. Size also limits a system's ability to carry equipment, and it is not common for these ROVs to have more equipment than a video camera. Low-cost ROVs should therefore be used only in the early stages of an archaeological investigation, in modest depths to confirm the identity of a site with the camera.

Larger observation-class ROVs have additional payload options and are capable of carrying still cameras, several video cameras, sonar systems, and so forth. These ROV systems have more power and can be used in deeper water. The size of these systems and modest power still limit the system capabilities, especially stability and the ability to move in strong currents, but these ROV systems are well suited for documentation in moderate water depths. Being both cost effective and easy to operate, these systems can perform a variety of tasks.

Most observation ROVs have been designed for in situ observation in moderate depths and have minimal collection capabilities, but these can be greatly increased by introducing a few modifications. For instance, since most of these vehicles are constructed from plastics, points of attachment for collection devices are mostly lacking. A skid can be built and used as a mounting bar, making it possible to attach simple tools like a one-function manipulator arm, a paired parallel laser device for taking measurements, additional cameras, scanning sonar, or corers. Synthetic foam flotation must be added to counterbalance the weight of these items. Typical thruster configurations of small ROVs lead to propeller wash, which disturbs sediments, reduces visibility, and disrupts the local area around the vehicle. To minimize these impacts, the ROV should be slightly positive during operations.

The depth limitation of observation ROVs may be overcome by using a tether management system in which the ROV is deployed in a "garage" or cage. The depth limitation is the result of a combination of limited thruster power and the drag force on long lengths of tether. With the garage, an ROV can be lowered to a few meters above the seabed close to the working area. The heavy garage hangs taut and stable and usually moves only a few meters. Signals to and from the ROV travel through the armored lifting umbilical to the garage. The ROV is then released from the garage to travel the short distance to the work site. The ROV tether is housed on a drum in the garage. The operator can operate the drum remotely to pay out or reel in the tether. The necessary tether has therefore effectively been reduced to less than 100 m and the drag force on the ROV tether has been mainly eliminated.

By using a tether management system, an observation ROV can operate in much deeper water, although this solution drastically increases both the weight and complexity of the ROV system and possibly requires a larger ship. The garage also protects the ROV as it passes through the splash zone and reduces the risks of an ROV operation. A free-swimming ROV increases the possibility of the tether becoming entangled in the thrusters of the ship or around obstacles on the seabed. The risks also increase in deep water because of the longer tether lengths and response time.

An alternative solution is downweights, which also increase the operational performance of observation-

Observation-class ROV (Sperre AS)

class ROVs (Sprunk et al., 1992). A downweight reduces the drag force from the tether; the tether is connected to the downweight from the surface ship, so the ROV system has to handle only the short tether from the downweight to the ROV. This arrangement can clearly increase the depth capability of small ROVs, but it also complicates the operation, for the downweight must be moved every time the ROV is moved to a new area on the seabed. The downweight can also be used as a sample receptacle. With small containers in the downweight, the ROV can collect several samples with a simple manipulator arm without having to surface every time a sample is taken. Other sampling hardware can also be added to the downweight, including suction sampling with a flexible hose operated by the ROV.

In a U.S. survey, several small observation ROVs were tested and compared to a diver's ability to inspect shipwrecks (Ryther et al., 1991). Seven shipwrecks in shallow water were video-documented, and accurate position and depth information was obtained for each site. The conclusion was that a properly equipped ROV system—with a camera, scanning sonar, positioning system, altimeter, and the like—would be capable of performing the required operations within a defined standard and schedule. (It should be noted that the purpose of these investigations was a channel clearance operation, not archaeology.) An average of 142 minutes was required for divers to chart the wrecks, compared to 219 minutes for the ROV. It was also suggested that with a properly equipped ROV and a trained and experienced crew, the ROV could have used the same or, in some cases, less time than the divers. This shows that, with proper planning and integration of sensors on the ROV system, an ROV can be more effective than divers, for it can collect data with fewer errors and greater consistency. As long as a task can be planned, there is no reason ROV systems cannot perform it as well as divers.

Work-class ROVs

When more advanced tasks are required, the size of the ROV and its power system must be increased. The first alternative is the small work-class vehicles. These ROV systems are larger and stable enough to carry tools for advanced documentation tasks but are also capable of light work operations including manipulative tasks. The

Supporter work-class ROV system on board *Edda Fauna* (right) and control room (below) (Kystdesign AS)

ROV industry has recently introduced a new generation of such ROVs for operation down to 3,000 m. These ROVs are perhaps the most interesting system for deepwater archaeological institutions, thanks to their high capability and modest price.

The next alternative is the medium and heavy work-class ROVs, which are often complex machines. Because of their size and the weight of the equipment fitted to them, manufacturers use hydraulics as their main source of power. The basic hydraulic system requires a pump to develop oil pressure and valves to control the flow of oil to the motor or ram, which in turn carries out the work required by the system. Hydraulic systems have several advantages over electrical systems, with a very good power-to-size relationship. These systems can be operated in fairly deep water without a tether management system and have been used down to 3,000 m without a garage but with a cable winch. A tether management system does, however, always increase work capability. To operate these large ROV systems, it is necessary to use specialized crews and to have a large backup system to deal with the complex operation. It is also necessary to use a larger ship to handle the substantial system weight of typically tens of tons and the power requirements (hundreds of kilowatts) of these systems.

Some large scientific and military ROV systems use a depressor instead of a tether management system to reduce the movement of long tethers. The depressor used by many research ROVs keeps the umbilical taut and near vertical so that umbilical drag is no burden to the ROV. The depressor also provides lights and overhead video images.

Power requirements dictate tether diameter, which in turn depends on tether dimensions and cable length. Cable drag is the most important factor limiting deepwater operations. It is mainly the introduction of the fiber optic tether cable and new materials that has made possible the large increase in ROV depth capabilities. The smaller diameter cable reduces drag, and data transfer speeds are very high with limited line loss. Most work-class ROV systems operate down to 3,000 m, while some systems are now capable of diving to 6,000 m.

The drawback of the large work-class ROV systems, apart from their complexity, is cost. The purchase cost is high, as is the operating cost, including ship and personnel. It therefore goes without saying that these ROVs are not used extensively in archaeological projects. In some cases, however, archaeological institutions have borrowed larger ROV systems from third-party organizations. In France and the United States, successful applications for funding can make it possible to borrow equipment from the national research laboratories, and in Norway oil companies have loaned equipment to cultural resource management projects related to oil and gas exploration in deep water. Some salvage operations have also had the necessary funding to use large work-class ROV systems.

Manned Submersibles

Manned submersibles—defined as any subsea vehicle that has a 1 atm cabin for human occupancy and is independent of a surface support vessel—have fascinated society for hundreds of years and played an important role in history by opening up the ocean depths to human exploration. The commercial and scientific utilization of manned submersibles reached a zenith in the late 1960s and early 1970s when companies such as General Dynamics, Rockwell, General Motors, and Westinghouse were actively involved in vehicle construction. In the past ten years manned submersibles have been largely supplanted by ROVs for many work-related tasks, principally because of the costs of insurance, manning requirements, and vehicle and handling system complexity. Manned submersibles are, however, still popular among many scientists, who feel that there is no substitute for direct, in situ observation of the marine environment. Of the estimated 160 commercial and scientific manned submersibles built in the past forty years, approximately forty are still operating. The majority can dive only to a few hundred meters. There are just five submersibles in the world that can reach 6,000 m. The Chinese and Russians are currently building new 6,000 m–rated manned submersibles. In 1960 the manned submersible *Trieste II* made history by descending 10,911 m in the Pacific Ocean to the bottom of the Mariana Trench, the deepest known point in any ocean (Thurman, 2003).

Manned submersibles have certain inherent technological characteristics that limit their ultimate efficiency. Manned presence requires the vehicle to be large and expensive for reasons of life support, and safety considerations require an increasingly complex certification effort. A typical manned vehicle weighing 10–20 tons also requires a large, expensive support ship and a sophisticated handling system. Moreover, only one or two scientists can participate on each dive. The available space inside the pressure sphere is also limited, which greatly reduces the supporting documentation a scientist can carry as well as instrumentation for data acquisition and analysis. Manned submersibles have therefore seen limited use in deepwater archaeology, except in the United States where various systems have been used in the Gulf of Mexico and the Great Lakes.

It is also possible to use atmospheric diving systems for dives to about 400 m. In a 1 atm suit, the diver is not exposed to the outside pressure and avoids decompression. These suits have legs and arms, and oil-filled, free-floating joints of various sizes provide a range of dexterity equal to 75 percent of an ambient diver. By virtue of these joints, the diver can kneel, lie flat, and generally orient himself to access most areas and perform most tasks. With a purpose-built thruster pack these suits can achieve midwater mobility and operating characteristics fairly close to that of soft-suited divers. Armored suits were, for instance, used to recover cargo from the wreck of the SS *Egypt* at 130 m in 1930, but they have seen limited use in deepwater archaeology projects (Pickford, 1993).

The U.S. Navy's nuclear-powered research submarine NR-1 was constructed by General Dynamics and launched on January 25, 1969. The 44 m long submarine was the first deep-submergence vehicle to utilize nuclear power. It has performed underwater search and recovery, oceanographic research missions, and installation and maintenance of underwater equipment to depths in excess of 800 m. Although it has now been decommissioned, NR-1 is probably the best-known submarine/submersible to be used for archaeological research, which is ironic since it spent the vast majority of its existence conducting covert operations.

The top speed of NR-1 was 4–6 knots on the surface, and it never strayed far from its support ship, although its nuclear propulsion provided essentially unlimited submerged endurance. NR-1 could travel submerged at approximately 4 knots for long periods, limited only by its supplies. It could study and map the ocean bottom, including temperature, currents, and other information for military, commercial, and scientific use. Precision maneuvering in proximity to the seafloor was provided by four ducted thrusters, two in the front and two in the rear. The vehicle also had planes mounted on the sail and a conventional rudder. It was also equipped with three 4-inch viewports on its bottom, nineteen 250 W gas dis-

Nuclear-powered research submarine *NR-1* (U.S. Navy)

charge lights, and eight 1000 W and two 500 W incandescent lights. It sported sixteen different low-light television cameras in various locations.

NR-1 had sophisticated electronics that aided navigation, communications, and object location and identification. It could maneuver or hold a steady position on or close to the seabed or underwater ridges, detect and identify objects at a considerable distance, and lift objects off the ocean floor. The submarine had no radar for surface navigation, but it did have a sensitive sonar system. Picking up objects from the ocean floor was an NR-1 specialty. With a hydraulically powered manipulator arm attached to its bow, it could pick up objects weighing up to a ton. The manipulator could be fitted with various gripping and cutting tools and a work basket that could be used in conjunction with the manipulator to deposit or recover items in the sea. Two retractable rubber-tired extendible bottoming wheels provided a fixed distance between the keel and the seabed, making the manipulator ability even more practical.

In the course of its many missions, NR-1 discovered numerous modern and ancient shipwrecks. It was also used by various scientific institutions to search specifically for, and to study at length, archaeological sites in the Atlantic Ocean and Mediterranean Sea. In 1995, NR-1 discovered the wreck of HMS *Britannic*, sister ship to the RMS *Titanic* on the White Star Line. During World War I, the *Britannic* served for a year as a hospital ship, but it was sunk in the Aegean Sea by a mine on November 21, 1916, near the island of Kea.

An ocean away, NR-1 also periodically visited the modern wreck of the passenger liner *Andrea Doria* to test its sonar imaging systems. The *Andrea Doria* was sunk in a collision with the M/V *Stockholm* near Nantucket Island, off the coast of Massachusetts, on July 25, 1956.

In 1995, while in the Mediterranean Sea, NR-1 conducted detailed reconnaissance on Skerki Bank, between Tunisia and Sicily. In 1988, R. D. Ballard led a research mission with the Wood's Hole Oceanographic Institution's (WHOI) Argo towed search system, which yielded numerous potential ancient shipwreck locations. Utilizing its impressive obstacle-avoidance sonar, sidescan sonar, and video systems, NR-1 was able to conduct a detailed study of the seafloor in the area identified with Argo nearly a decade before. NR-1 returned to the Mediterranean Sea in 1997 and supported further research at Skerki Bank with WHOI and Ballard, resulting in the location and mapping of numerous well-preserved, ancient, medieval, and modern shipwrecks. This project was a watershed event for the development of deepwater archaeology and also for the development of multidisciplinary academic, corporate, and governmental/military collaboration for deepwater archaeological research in general.

Also in 1997, while searching for the lost Israeli submarine *Dakar*, NR-1 located two nearly pristine shipwrecks from the Iron Age along the Egyptian coast. In 1999 these important shipwrecks, carrying mostly amphoras, were mapped in with the ROV *Jason* from WHOI. Unfortunately, these wrecks have yet to be excavated or studied further.

In the year 2000, NR-1 returned to the Mediterranean Sea and made time available for both geological and archaeological research in the Ionian Sea. In concert with ProMare, a nonprofit marine research and exploration corporation, NR-1 surveyed the central Ionian Sea between Italy and Greece, south of the Straits of Otranto, for a period of five days. In addition to discovering a vast field of lithoherms (ancient, deepwater coral reefs), the submarine located one ancient shipwreck that was buried almost completely, with only the upper edges of the starboard side of the hull exposed, along with one iron anchor and several small pottery vessels.

While working along the east coast of the United States, NR-1 assisted NOAA in its study of the wreck of the USS *Monitor*, a 987-ton armored turret warship. Caught in a storm off Cape Hatteras, it foundered on December 31, 1862. The wreck was discovered in 1973 and is now a national marine sanctuary. Major components of its structure have recently been recovered, to be followed by extensive conservation. In 2007 the USS Monitor Center opened in Virginia, displaying and interpreting the site.

In 2002, NR-1 supported extensive scientific research in the Gulf of Mexico, including the investigation of seemingly ubiquitous biological and geological seafloor formations known as chemosynthetic communities. The submarine also supported the archaeological investigation of a shipwreck in the Mississippi Canyon region of the Gulf of Mexico, approximately fifty miles south of New Orleans and at a depth of 800 m. The copper-clad shipwreck is 20 m long and sits upright on the seafloor. It was discovered accidentally after a petroleum exploration corporation inadvertently placed an 8-inch natural gas pipeline through the midships section of the wreck, nearly bisecting it.

Also in 2002, NR-1 assisted in the mapping of the wreckage of the rigid airship USS *Akron* (ZRS-4). The *Akron*, first of a class of two 6,500,000 ft^3 rigid airships, was built in Akron, Ohio. Commissioned in late October 1931, the airship spent virtually all of its short career on technical and operational development tasks, exploring the potential of the rigid airship as a naval weapons system. While beginning a trip to the New England area, Akron encountered a violent storm over the New Jersey coast and, shortly after midnight on April 4, 1933, crashed tail-first into the sea. Much of its collapsed framework remains visible on the continental shelf, nearly seventy years after the great dirigible went down.

NR-1 has now been deactivated, ending its use in deepwater archaeological research projects. Other minisubmarines will not be able to replace its capabilities, although submarines will probably be used on future marine archaeological projects, including ProMare's new range of scientific submarines.

The Basic Tools

Hugin AUV (above and left) (Kongsberg Maritime)

Autonomous Underwater Vehicles

ROVs employ an umbilical cable to carry both power and telemetry from a mother ship to the vehicle. A growing number of research vehicles, the so-called autonomous underwater vehicles (AUVs), operate without an umbilical tether. AUVs are increasingly used to locate and document shipwrecks (Bellingham, 1992; Mindell and Bingham, 2001; Conte et al., 2007). They have a clear competitive advantage over their tethered relative the ROV and also over towed equipment. The free-swimming AUV is not dependent on a long and expensive cable, removing the need for large winch systems and ships with position-holding capability. AUVs rely on either a completely autonomous operation, where the system is preprogrammed to carry out specific tasks and then surface with the results, or untethered (UUV) mode, with a man-in-the-loop for mission control. AUVs are now reaching a threshold of cost and demonstrated capability that encourages their use to survey the seabed and look for archaeological sites (Størkersen and Indreeide, 1997; Chance et al., 2000).

To search for archaeological sites on the seabed, an AUV can be equipped with lights and video cameras, sidescan sonar, magnetometer, sub-bottom profiler, and other equipment. It can cruise at a speed of at least 3 knots along a set of predefined tracklines. Since the battery capacity is more than fifty hours, it can stay submerged for several days at a time. Positioning can be achieved using an acoustic underwater positioning system or by surfacing the AUV at regular intervals and using GPS buoys to adjust position. In addition, Doppler sensors and other motion reference units are used to make time-of-flight measurements to increase positioning accuracy. Knowing its exact position at any time, an AUV can follow the established search pattern while its sensors collect the required information. On completion, the vehicle surfaces and the results can be interpreted.

Practical experience with state-of-the-art survey ROVs indicates a maximum speed capability of 3 knots in very shallow water, decreasing to less than a knot at 400 m. This reduction in speed is automatically transformed into higher costs. An AUV can maintain its speed in both shallow and deep water, since there is no cable drag. AUVs will become far more versatile as the technology advances. Hybrid versions are being developed that will not only survey the ocean floor but also hover to conduct inspections and collect artifacts. Several projects have made use of AUVs, and more will follow when availability increases and cost decreases.

The Danish AUV *Maridan* carried out a research project in cooperation with the National Museum of Denmark's Centre for Maritime Archaeology in the late 1990s. The project tested sensors and payload instrumentation, such as sidescan sonar, a sub-bottom profiler, and a digital camera to search for medieval shipwrecks, Stone Age settlements, and sites from the Viking period in the waters off the Danish and Swedish coasts. This was the first time an AUV was used for underwater archaeology.

AUVs from WHOI have also been used to document shipwrecks in relatively deep water. In 2005 an AUV from WHOI was used to document (including microbathymetric data and photomosaicking) a shipwreck in Greece.

C&C Technologies was one of the first companies to use an AUV to map and survey large areas of the seafloor, especially in connection with oil and gas projects in the Gulf of Mexico. It has shown that it is possible to scan the seafloor at much greater depths than previously thought possible, except at exorbitantly high cost with a deep tow system. These surveys have already located several shipwreck sites, including historical shipwrecks in the Gulf of Mexico and the Mediterranean. Other discoveries to date include the *U-166*, the only German submarine believed to be sunk in the Gulf of Mexico, and the *Ark Royal* off Gibraltar.

Limited Economic Resources

Marine archaeology is not characterized by large budgets. Unfortunately, the cost of using the equipment and ships required to do deepwater archaeology is high. The few deepwater projects accomplished so far have been able to utilize equipment from oceanographic institutions or found a sponsor. In some cases the ship's cargo has been valuable enough for individuals or companies to invest in the salvage of the cargo. Other sources have included government programs and national cultural resource management programs, which state that commercial projects in the deep sea (e.g., oil and gas) have to pay for the archaeological investigations. Most ROV vessels with crew cost at least US$50,000 per day. This represents perhaps the biggest obstacle for the development of deepwater archaeology.

THREE

History of Deepwater Archaeology

Over the past thirty years scientific, commercial, and military projects have located several archaeological sites in deep water. One of the first shipwrecks discovered in very deep water was an ancient amphora carrier found at a depth of 760 m during a closed-circuit television search for the wreckage of a commercial airliner off the island of Elba in the Mediterranean in 1952. Two years later, during a pipeline route survey from France to Africa, twenty-four intact ancient wrecks were supposed to have been located. In 1966, while searching for a hydrogen bomb lost when two planes collided off the Spanish coast, the crew of the submersible *Aluminaut* reported finding two intact ships at 610 m, their masts still standing and the cannon projecting from the gunports (Marx, 1990). Oceaneering International, one of the largest subsea companies in the world, discovered and documented an ancient (AD 500) shipwreck laden with amphoras in the Tyrrhenian Sea at 3,453 m using their 7,000 m ROV *Magellan*, which was purpose-built for object search and recovery operations. These early reports showed that the possibility of finding deepwater shipwrecks was high and also fueled the theory of the perfect preservation conditions in deep water.

Since then, several archaeological institutions and salvage companies have used remote sensing equipment to locate and investigate deepwater archaeological sites. However, this activity has with a few exemptions been sporadic, and only a few institutions and companies have been able to work systematically over a long period because of the high complexity and high cost.

The application of underwater technology in marine archaeology is now common practice in United States and Europe. In Africa, Australia, Asia, and South America only a few deepwater projects have been initiated. The Mexican government hired a Russian oceanographic vessel to locate the remains of a Spanish galleon in deep water in the Gulf of Mexico, but it failed to locate the wreck. The plan was to use three Russian submersibles to salvage the wreck.

In general, the majority of projects carried out so far have typically been pure observation projects in which sites have been located and then documented using an ROV. Advanced archaeological tasks have typically not been carried out, although a handful of projects have included more advanced tasks. In this chapter I review the most prominent research projects, organizing them geographically, and follow up with a look at several commercial, essentially non-research projects—which nonetheless also illustrate many aspects of the tools and methods developed for deepwater archaeology.

North American Waters

One of the first to use the emerging technologies to locate and investigate deepwater archaeological wrecks was the American oceanographer Willard Bascom. Bascom was of the opinion that thus far unknown archaeological data could be obtained from well-preserved deepwater wrecks. Spurred by a lifelong interest in naval history and ancient ships, in 1971 Bascom designed a research vessel, called the *Alcoa Seaprobe*, to locate and raise deepwater wrecks (Bascom, 1976).

The *Seaprobe* was equipped with a dynamic positioning (DP) system, making the vessel capable of holding any position without anchoring. By using several propellers capable of producing thrust in any direction, the

Alcoa Seaprobe (Willard Bascom)

Recovery pod and arm for the *Alcoa Seaprobe* (Willard Bascom)

system maintained the ship's position above a site in spite of winds and currents and made it follow selected search patterns. At the center of the *Seaprobe* was a derrick, similar to those on a drillship, about 40 m tall. Beneath the derrick was a center well through which the drill pipe could be lowered. The *Seaprobe* had equipment to handle 21 m of pipe at a time and was capable of handling 400 tons of weight (pipe and load), which gave a maximum pipe length of 3,000 m. The lower end of the pipe was weighted with 20 tons extra weight to keep the pipe taut and nearly vertical while the ship was moving.

By mounting instruments at the lower end of the pipe, the *Seaprobe* could be used to find and recover wreck sites in deep water, at least in theory. The drill pipe had several advantages. It did not flex much and could be used to transmit rotational force to the bottom, and water could be sent down to run turbines at the bottom for power or to dust off objects on the site. A pin could be extended from the lower tip to steady the instruments and make delicate recovery operations possible.

To search for sites, a search pod was attached to the bottom of the drill pipe. This pod was equipped with sonar, cameras, and lights. Signals to and from the pod were transmitted through a cable attached to the outside of the pipe. Two sonar systems were used simultaneously; one scanned directly forward to prevent the pod from running into obstructions, while sidescan sonar scanned 300–400 m to each side of the pod to locate possible sites. The ship used an array of base stations (radio waves) on land to navigate precisely. Alternatively, an array of buoys with radar transponders was established to create a navigational network on the ship's radar. The pod was lowered to the bottom by extending the drill pipe, while the ship moved along predefined search lines. If something was detected, the pod was lowered even farther for a closer scan with the sonar and a direct examination with the cameras.

After a target had been selected for intensive study and possible recovery, the search pod was to be retrieved and replaced with an examination-recovery pod (which was, however, never built). The pod would consist of a set of tongs or grab buckets mounted on an arm attached to the bottom of the pipe. Guided by the cameras, these could be used to lift objects, which would then be recovered by raising the pipe to the surface. Water jets could be used to propel or wash sediments away from objects. A separate duster, a ducted propeller mounted on a tripod placed above the site, was also designed to remove sediments from shipwrecks.

Bascom was also the first to introduce deep-towed sleds equipped with low-light cameras and sonar to search for deepwater archaeological sites. These sleds were towed along the end of a long cable, only few meters above the seabed, positioned by an underwater positioning system.

Bascom also designed a jet rake to locate objects hidden in the sediments; this was a sieve that could be towed behind the ship and dragged through a muddy bottom to strain out buried artifacts. He also designed a huge set of tongs that could be used to enclose a wreck and some of the underlying mud and to raise it almost to the surface, where it could be worked on by more traditional means. These tongs had pontoons with adjustable buoyancy. Neither the jet rake nor the tongs were ever realized.

Bascom used another unit, the TVSS (television search and salvage system), to salvage objects from deep water. This unit was lowered to inspect a site via cameras and also had sonar and a hydraulic grab bucket capable of lifting up to five tons. It was used to salvage items from a Boeing 727 airplane that went down in Santa Monica Bay, California, in 1969, in 290 m depths (Bascom, 1991). To shift the position of the cage, a winch pulled in on a system of anchored ropes. Many salvage companies still use this system, but with propellers to move the cage around (Crawford, 1995). This tool can, however, severely damage a site, and it should not be used for archaeology. Although some of Bascom's ideas were naive and would have destroyed archaeological sites if they had been used, some ideas have been modified and used with success. The idea of raising intact wrecks to excavate them in shallow water has, for instance, been proposed by several archaeologists in recent years.

USS Monitor

On December 31, 1862, the USS *Monitor*, the famed ironclad ship of the U.S. Navy, sank at sea during a storm as it was being towed from Virginia to South Carolina. The vessel played an important role in the development of naval technology and warfare and signaled the end of the era of wooden sail-powered ships as the citadel of sea power. The *Monitor*'s unique turreted design was the first comprehensive response to the technological innovations that would revolutionize both naval architecture and warfare at sea during the nineteenth century.

In 1973 the heavily damaged remains of the *Monitor* were located, approximately 25 km southeast of Cape Hatteras, and in 1975 the site was declared a federally protected marine sanctuary, the first of its kind in the United States. The ship was found with a 100 kHz EG&G side-scan sonar device in combination with a Varian proton magnetometer used to identify and assess the magnetic

The ironclad USS Monitor (Mariners' Museum)

signatures of the sites identified by the sonar to reduce the number of targets (Watts, 1987).

In 1974, the *Alcoa Seaprobe* was used to investigate this site. The search pod was attached to the drill pipe and lowered to the site, more than 70 m below. The cameras on the search pod took more than 1,200 35 mm photographs and several hours of videotape. Selected photos were used to construct a photomosaic of the entire wreck. After this survey a series of investigations of the wreck were carried out. In 1977 the manned submersible *Johnson Sea Link* from the Harbor Branch Foundation was used to do a photogrammetry survey of the site and recover selected materials for testing and analysis. After a series of dives to examine the wreck and familiarize the pilots, a dive team was locked out of the submersible to install a baseline designed to control the photogrammetric data collection. After a baseline had been installed, three sets of stereo photos were taken with the submersible, and some objects were recovered by the divers.

The *Johnson Sea Link* was used on the site again in 1979. A series of permanent provenience stations were established at the perimeters of the site. Divers from the submersible carried out a small excavation using a hydraulic dredge powered by a centrifugal pump, operated by one of the submersible's thruster motors. During the excavation, a 35 mm camera was used to take stereoscopic photographs of objects exposed in situ, while elevation within the excavation was controlled by a bubble level. Additional documentation with video and photography was done, and several more artifacts were recovered (Watts, 1987). A third investigation using the *Johnson Sea Link* was carried out in 1983. In 1985 a bathymetric map, sub-bottom profiling map (to identify objects buried in the sediments), magnetic contours map, as well as additional sidescan sonar images were collected to clearly define the extent of the site.

The ship was also investigated in 1987. Unlike the previous *Monitor* expeditions, all data were collected by an ROV, the navy's Deep Drone System operated by Eastport International, from the USNS *Apache.* Comprehensive readings determined the extent of corrosion of the wreck; precision photomosaic, intensive artifact, and engineering and structure surveys were conducted; and three-dimensional modeling of the wreck was developed from sonar information.

Essential to this operation was Eastport's ALLNAV integrated navigation system, which used GPS or microwave positioning to position the ship and an acoustic underwater positioning system to position the ROV. Knowing exactly where the data come from is a key component of gathering archaeological data. The precise position data collected at the *Monitor* site enabled the hundreds of 35 mm and 70 mm still photos taken at the site to form a complete and accurate photomosaic of the hull in plan view and elevation view. Video was obtained via fiber optic cable, but the ROV was not able to investigate the complete site because of the strong currents. Three-dimensional acoustic imaging was collected with a downward-looking ROV-mounted sonar transponder. The sonar data, together with navigation information, were processed and used to create a color depth contour map, three-dimensional shaded perspectives of the wreck, and animated flyarounds. The survey also discovered that the *Monitor* continues to deteriorate and corrode.

Scuba divers were used for the first time to investigate the *Monitor* in 1990 to record parts of the site accessible only to free-swimming divers. The divers used video to document the site. By digitizing the video images and using data analysis software, it was later possible to acquire accurate measurements from the video images. A computer captured images from the video source via a digitization board or a frame grabber and converted the analogue video signal to a digital, high-resolution computer format. Measurements from the digitized images were made with BioScan's Optimas data analysis software and later used to construct a complete site plan (Farb, 1992).

NOAA determined that the collapse of the *Monitor*'s hull was imminent and would result in the loss of much of the ship's structure and contents. To head off this threat, NOAA and the U.S. Navy conducted missions to the sanctuary in 1993, 1995, 1998, 1999, 2000, and 2001 to survey the wreck's condition, stabilize the hull, and recover significant components and artifacts, in accordance with NOAA's comprehensive preservation plan for the *Monitor* (Alberg et al., 2008).

In 2001, U.S. Navy divers recovered the *Monitor*'s innovative steam engine and a section of its hull. In 2002 the project team successfully removed a section of the armor belt and hull. Finally, the turret was partially excavated, rigged, and recovered with all its contents. These items are currently undergoing conservation and now form an impressive museum display.

The *New Jersey* Project

On the evening of February 25, 1870, the 494-ton round-sterned, wooden steam freighter *New Jersey* departed Baltimore Harbor bound for Norfolk, Virginia. At about midnight, fire was discovered amidships and between decks, and the crew was forced to abandon ship. In 1975, Nautical Archaeology Associates, asked by a bay waterman to examine a large obstruction not noted on any nautical charts, discovered the shipwreck at only 25 m depth. No institution was prepared to conduct an intensive survey of the site because of its enormous scale (estimated 2.5–6 million artifacts), depth, and low visibility. The site was, however, inspected in 1985 using sidescan sonar, and in 1986 National Geographic used a Mini Ranger ROV to produce video and photographic records of the site. National Geographic wanted to use this site as a testing ground in their plans to use ROVs for deepwater archaeological investigation (Shomette, 1988).

Thus, in 1987 the *New Jersey* became the object of one of the earliest intensive experiments in robotic and underwater archaeological survey. Researchers from National Geographic, Deep Sea Systems International, the U.S. Navy, and Woods Hole Deep Submergence Laboratory used and evaluated several different systems believed suitable for archaeological investigation to study the *New Jersey*. The project team had a fleet of three ROVs equipped with high-resolution low-light cameras, an ultrasonic high-accuracy ranging and positioning system, scanning sonar, and several additional tracking and recording systems. Visibility never exceeded a meter, so video work proved impossible, and navigation of the three ROVs had to be entirely by sonar and onboard instrumentation. The *New Jersey* project was one of the first attempts to show that archaeological investigation could be done with ROVs and various remote sensing equipment under water.

The *Hamilton* and *Scourge*

The Anglo-American War of 1812 involved several naval actions on the Great Lakes but was not responsible for the sinking of two schooners, *Hamilton* and *Scourge,* that sank in a storm in 1813 while patrolling Lake Ontario. They were discovered in 1973 by an officially commissioned Canadian research project using sidescan sonar. Deep and cold fresh water in the lake offers excellent preservation conditions, and the ships are in an extraordinarily good state of preservation, standing upright on the seabed. The two ships have been the target of several investigations (Cain, 1991).

Hamilton and *Scourge* lie at more than 90 m. They were investigated by the Hamilton-Scourge Foundation/National Geographic Society in 1982 (Nelson, 1983). This investigation revealed that the ships are basically intact with a full range of artifacts. The guns can be seen on deck, three of the four masts are still standing upright, and boarding axes are in their racks. The visibility is poor, but an ROV named RPV, designed by the American company Benthos, took 1,500 still images and 23 hours of video during the six-day operation. The two schooners are archaeological treasures of international importance, and the two ships and their artifacts provide a blueprint for naval practices of that period. There are detailed plans to raise and display the two vessels in a lakeside museum, and various expeditions have used ROVs to film and document the shipwrecks on the lake floor.

Another archaeological survey of *Hamilton* and *Scourge* was carried out by National Geographic in 1990 utilizing WHOI's ROV *Jason.* Using Kongsberg scanning sonar mounted on *Jason,* a three-dimensional characterization of the warships could be made. A black-and-white still camera, mounted in vertical and horizontal position, was also used to make a series of controlled photo runs. The photos later underwent enhancement to remove particle backscatter and to compensate for uneven lighting and were finally used to create photomosaics of the wreck sites. This work was based on only one initial photo. Starting with this and using several tie points, a new image was planned, zoomed, and blended to fit the initial image. A full mosaic this one site consists of around one hundred photographs selected from more than four thousand taken.

The *City of Ainsworth*

In 1990 the sternwheeler *City of Ainsworth,* which sank in 1898, was discovered in 111 m in Kootenay Lake, British Columbia, with sidescan sonar. The wreck site was inspected with a small ROV, equipped with scanning sonar, and found to be mainly intact (Beasley, 1991).

Deepwater Archaeology in the Gulf of Mexico

It is well known that the Gulf of Mexico contains many shipwrecks of archaeological and historical significance dating back to the discovery of the New World. Some have been discovered and documented, yet the majority remain unstudied. In addition to discoveries as a result of exploration activities by universities and research institutes, many important discoveries have been made in recent years during the exploration, development, and production of oil and gas resources; it is estimated that between three hundred and four hundred shipwrecks have been discovered during oil- and gas-related activities. Historical research conducted for Minerals Management Services (MMS), the federal agency that manages offshore oil and gas development, indicates that more than four hundred ships have sunk on the federal Outer Continental Shelf (OCS) dating from 1625 to 1951; thousands more have sunk closer to shore in state waters during the same period. Only a handful of these have been scientifically examined by archaeologists (Irion, 2001).

MMS is required to ensure that activities it funds or permits, such as lease sales, the drilling of oil and gas wells, and the construction of pipelines, do not adversely affect significant archaeological sites on the OCS. To assess the potential to affect archaeological resources by proposed oil and gas activities, MMS has funded archaeology studies to ascertain where on the OCS these sites are likely to be. MMS reviews nearly two thousand planned wells and pipelines every year for their potential effect on archaeological sites on the OCS.

Archaeological sites on the OCS are most likely to be either prehistoric Native American sites dating from the end of the last Ice Age, when sea levels were about 70 m lower than they are today, or historic shipwrecks. In areas where archaeological sites are likely to be found, the oil and gas industry is required to conduct surveys of the seafloor using remote sensing instruments. The data collected are reviewed by archaeologists, who write reports on their findings for submission to MMS. MMS archaeologists, in turn, use these reports to review applications from industry to drill wells or construct pipelines. Several new shipwrecks sites have been discovered as a result of these surveys conducted by the oil and gas industry (Warren et al., 2007).

The Steam Yacht Anona   The discovery of the wreck of the steam yacht *Anona* was reported to MMS in July 2002 by BP Exploration after an examination of the site by C&C Technologies. The shipwreck lies in 1,200 m of water in the Viosca Knoll area and was identified by a combination of remote sensing instruments packed into C&C's Hugin AUV and cameras mounted on an ROV.

The *Anona* was a 117-foot, steel-hulled, propeller-driven steam yacht built for wealthy Detroit industrialist Theodore DeLong Buhl in 1904 by the well-known yard of George Lawley and Sons in Boston. *Anona* met a somewhat ignominious end in 1944 when it sank in the Gulf of Mexico carrying potatoes to the British West Indies after the plates under the steam engine buckled. The nine-man crew was rescued by three PBY planes after spending two days adrift in a raft.

The wreck of the *Anona* is in an excellent state of preservation, sitting upright on its keel. It is buried in the bottom to just above what would have been the vessel's waterline, but the deck is relatively clear of sediment. Most of the wooden decking has deteriorated, but the steel hull and machinery appear to be in good condition. Once the setting of glittering soirees for Detroit's social elite, it now is home only to countless anemones and crabs in a world of eternal darkness.

Copper-sheathed Shipwreck   A copper-clad ship was discovered accidentally after a petroleum exploration corporation inadvertently placed an 8-inch natural gas pipeline through the midships section of the wreck, nearly bisecting it. The ship lies in over 800 m of water in an area of the Gulf known as Mississippi Canyon. All that remains of the vessel is its lower hull, which is clad in copper sheathing and shows some evidence of burning. MMS archaeologists believe that a fire on board the ship may have been the cause of its sinking.

The ship is about 20 m long and its wooden hull is covered with thin copper sheets, a means of protecting ships

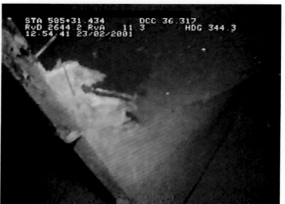

Copper-sheathed shipwreck (two views), Gulf of Mexico (ProMare)

from wood-eating marine organisms used by shipbuilders from the end of the 1700s to the mid-1800s. Since copper sheathing was quite expensive, it is unusual to find it on small merchant vessels, which has led MMS archaeologists to speculate that this could have been a naval vessel.

To investigate the ship in more detail, MMS entered into a cooperative agreement with Texas A&M University to conduct an archaeological investigation of the shipwreck. In 2002 the project utilized the *Carolyn Chouest* and the U.S. Navy research submarine NR-1 to investigate the shipwreck in combination with an ROV. The site was documented and selected artifacts recovered to the surface.

The *Western Empire*   The *Western Empire* was first reported to MMS as a possible shipwreck in a remote-sensing survey conducted for Shell Oil in the early 1980s. Its identification and historical significance were unknown until scientists from Texas A&M University visited the site as part of a study of another nearby shipwreck.

Lost September 18, 1876, en route from Liverpool to New Orleans, the *Western Empire* was carrying a load of lumber (explaining the lack of cargo at the wreck site) when it sprang a leak and foundered. Ten men drowned in the sinking. The ship is remarkably well preserved and heavily colonized with all manner of organisms, including anemone, coral, amphipods, mollusks, crustaceans, eels, and numerous species of fish. The 60 m long wreck rests in 400 m of water off the coast of Louisiana.

Archaeologists from Texas A&M University and MMS used remotely operated cameras and a one-man submersible to videotape the nearly intact sailing ship for later analysis. The cameras recorded large sections of the wooden hull standing over 3 m off the seabed and even parts of the rigging but, surprisingly, no evidence of cargo. MMS archaeologists were later able to use this information to identify the shipwreck.

The *Mardi Gras* Project   The *Mardi Gras* shipwreck site is off the Louisiana coast at about 1,220 m depth. The site was discovered during a pipeline survey. Preliminary studies of the site revealed a few large and several small diagnostic artifacts scattered on the silty, flat bottom in an area approximately 20 by 5 m. The largest artifacts were two anchors, one cannon, one camboose (ship's stove), and a wooden chest that contains what appears to be weaponry. Small artifacts included glass bottles, white porcelain, nautical instruments, small metal artifacts, and other ceramics. The remains of the ship's hull were visible in certain areas, but the majority of the ship appeared to have been eroded or buried under the sediment. The small size of the site suggests that these are the remains of a typical small commercial ship that frequented the gulf coast in the early nineteenth century. According to historical maps of that period, the wreck lies at a location where several major commercial shipping routes intersected.

The early nineteenth century is one of the most crucial but least studied periods in New World history and is a turning point in the history of the Gulf of Mexico and the nascent United States. The period between the addition of the gulf coast states into the Union (Louisiana in 1812, Mississippi in 1817, Alabama in 1819, Florida and Texas in 1845) and the American Civil war is the golden age for the Gulf of Mexico from a commercial point of view. Port registries indicate that most of this commerce was carried out on board small ships that frequented the gulf ports to load the cargoes that descended to the gulf through the Mississippi River system. Once larger international cargoes were accumulated, they were transported on ocean-going ships to reach European, and especially British, ports; but the burden of transportation within the gulf fell upon these small sailing ships represented by the *Mardi Gras* shipwreck site. A thorough understanding of this small craft and its cargo will significantly contribute to our historical picture and understanding of the economic development of the American south, the growth of the nation, and the dynamics of maritime trade in the Gulf of Mexico.

Texas A&M University was contracted to perform the marine archeological investigation at the site, document the remains, and recover selected artifacts from the surface. This work was carried out in the summer of 2007 (Ford et al., 2008). At this depth, all of the work was carried out by ROVs using a range of specialized equipment tailored for performing tasks under these conditions. The equipment was a combination of off-the-shelf vehicles and tooling adapted from the offshore oil and gas industry and specialized tooling specifically engineered for the needs of this project.

Veolia Environmental was contracted to provide the 256-foot *Toisa Vigilant* and a Perry Triton XLS-17 for the project. This ROV is a 150 hp hydraulic work-class ROV system. The project also utilized a Sperre Sub-fighter 7500 ROV constructed specifically for the project with tooling developed by NTNU for the Ormen Lange project. Special tools used on the *Mardi Gras* shipwreck project included high-resolution cameras and lights, feedback manipulator, standard manipulator, excavation dredge and screening system, scaling lasers, suction pickers to recover small artifacts, and scoops to recover small artifacts.

The project also developed special tooling for retrieval of large artifacts. Once deployed and lowered to the seafloor, the Triton ROV positions this grab-unit over the artifact to be retrieved, and an hydraulic connection is established to close the two grab halves. Although crude and not complying with archaeological standards, the recovery tool was deemed necessary to bring up large artifacts quickly. Clamshell-like devices are admittedly experimental and are employed only after careful consideration of several factors.

Basic positioning of the ROV was provided by a Kongsberg SSBL system, which has a typical accuracy of 0.5 percent of water depth. To minimize the SSBL error as

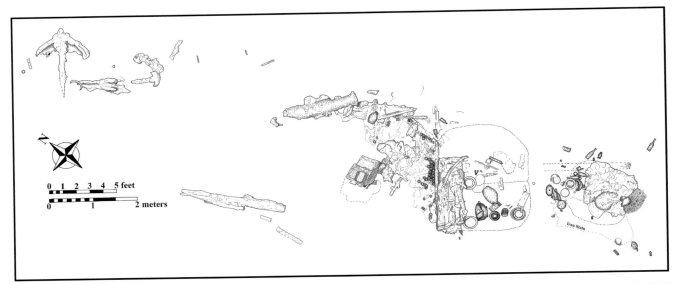

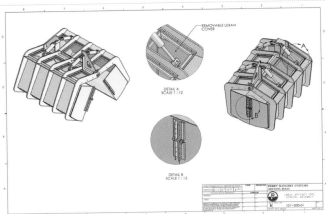

Mardi Gras shipwreck site plan (top); Triton XLS ROV used on the *Mardi Gras* project (left), equipped with two manipulators, suction picker, and suction dredge; and large artifact recovery tool (above) (B. Ford; Texas A&M University)

a function of the water depth (up to 6 m in this case), the team used corrections provided by a fixed transponder on the seabed as well as the Kongsberg high-accuracy inertial navigation system, or HAINS, providing a more stable and precise reading of the SSBL signal, which increased the accuracy of all measurements taken.

The first phase of the project involved multibeam bathymetry and sub-bottom profilers to produce a high-resolution, accurately positioned three-dimensional image of the seafloor and imaging of buried features, hull structure, or large artifacts to aid in planning the investigation.

Mapping was done by capturing high-resolution, orthogonally corrected photomosaics over the course of the project. To accomplish this, the Triton ROV was deployed with scanning sonar, high-resolution video cameras, and digital still cameras. Digital still and video imagery was acquired at an ROV flight altitude of 4 m, which provided swath coverage of 4.5 m. To achieve coverage with a 60 percent overlap, transect lines spaced approximately 3 m apart were flown.

A controlled recovery of artifacts exposed on the seafloor was also carried out. Excavation was limited to clearing areas around the large artifacts and features to aid their recovery. Site Recorder 4, developed by 3H Consulting, was selected as the main software system for recording, registering, and cataloging artifacts from fieldwork through the conservation process. Site Recorder 4 can manage thousands of artifacts, drawings, photographs, video clips, documents, and geophysical data files, which can be linked for analysis and interpretation. Artifacts are cataloged by their functional class and material of composition, photographed before and after recovery, plotted on the site map, and assigned a unique artifact record number.

The Conservation Research Laboratory at Texas A&M University, under the direction of Donny Hamilton, will conduct all conservation of recovered material. Methods will vary with the type and number of artifacts collected. Once conserved and properly labeled, artifacts will be delivered to the Division of Archaeology of the Louisiana Department of Culture, Recreation, and Tourism for permanent curation and future study.

French Atlantic and the Western Mediterranean Sea

The French have pioneered underwater exploration and have been responsible for such breakthroughs as the autonomous diving suits of Gagnan and Cousteau and the deepest manned submersible dive ever, of almost 11,000 m. This tradition is also present in the underwater archaeology work carried out in France with advanced programs for both shallow and deepwater archaeology.

Underwater archaeology is usually carried out by scientists and technicians of DRASSM (Department for Underwater Archaeological Research) and by a team of specialists from the national research centers, such as CNRS. Close to seven hundred shipwrecks, six hundred in the Mediterranean alone, have been recorded by these institutions, resulting in greater knowledge of maritime commerce and ancient ships in French waters.

DRASSM was founded in 1966 as a national service located in Marseilles and Annecy. Its aim is to manage all underwater archaeological activities and to implement laws governing maritime cultural goods. The department is also responsible for evaluation and inventory of maritime cultural material, preservation, research, publications, and exhibitions.

The area of authority is large and covers over 10,000 km of coastline. The area extends 12 nautical miles (22 km) outward from the shoreline, to span the surface of over 200,000 km². To survey this area the French government (Ministry of Culture) built a research vessel, the *Archéonaute*, for underwater archaeological work. The *Archéonaute* is 30 m long and can accommodate up to twelve divers with the necessary equipment for diving, a photo laboratory, and so forth.

French institutions typically use magnetometers to locate sites down to about 100 m, but they also rely on sidescan sonar and chance discoveries by divers. In deeper water no systematic surveys are done, and the few really deep wrecks known have been discovered by coincidence by others, such as the French Navy with its manned submersibles and COMEX.

French institutions like DRASSM have been involved in several deepwater archaeology projects, Some wreck sites at relatively great depths (60–80 m) have been explored with traditional diving techniques, even though this approach allows only elementary documentation, and under difficult and dangerous conditions.

Alternative solutions are therefore used to carry out methodical investigations in deep water. For example, a saturation diving operation was completed in 1988 on the *Héliopolis 2* wreck near Toulon. Four saturation divers from the French Navy investigated the site at almost 80 m. These divers produced a site plan of the shipwreck and raised more than twenty amphoras after tagging them manually.

In addition to deep diving, many deepwater archaeological sites have also been investigated with advanced underwater technology. French archaeological institutions have cooperated with IFREMER (French Research Institution for the Exploitation of the Sea) and used equipment from underwater technology companies such as COMEX in several projects. IFREMER, a government agency under the Ministry for Higher Education and Research, has the mission of directing, funding, and promoting ocean research and development. It has a fleet of twelve research vessels, two manned submersibles (*Nautile* and *Cyana*), ROVs, and other equipment. DRASSM and IFREMER have cooperated on many marine archaeological projects, realizing research that would not otherwise have been possible because of the complexity and cost—especially the *Arles IV* shipwreck site.

Arles IV Site

This shipwreck was discovered in October 1988 by the submersible *Cyana* at a depth of 662 m and investigated by DRASSM in March 1990. It is located in the Gulf of Lions around 40 nautical miles (74 km) from the mouth of the Rhône. In 1993 it was investigated by an archaeological expedition called Nautilion. The ship, originating in Bétique in the south of Spain, probably sank between AD 25 and 40. It has gradually deteriorated, and all that remains today is a vast (30 by 10 m) field of amphoras.

About 950 amphoras have been counted, but the height of the pile suggests that the amphoras are deposited in several layers and that the ship actually transported between one and two thousand of them. The cargo is arranged systematically: a large group of amphoras with pickling brine and preserved fish is in the center; oil amphoras are placed around this central portion, and each extremity is occupied by small, flat-bottomed wine amphoras; pottery and copper ingots are placed at the edges to complete the ship's cargo. To the north a leaden stock of an anchor and an iron anchor indicate the position of the prow. Some amphoras typical of the Balearic Islands region are deposited at the periphery of the site and indicate a port of call in that archipelago (Long, 1995).

The manned submersible *Cyana* was developed in the late 1960s by IFREMER and designed to function down to 3,000 m. The submersible has a crew of three and is manually piloted while traveling at or near the seabed. Its equipment includes a five-function manipulator arm with built-in tools, mobile basket (50 kg capacity) for samples, video and photo camera, data recording equipment (time, heading, depth, inclination, temperature, etc.), and scanning sonar.

Links with the operations room on board the support ship are maintained via a two-way UHF transceiver and, during dives, with an underwater telephone. In 1993 the site was investigated with another manned vehicle, the *Nautile*. This submersible can be used for observations and operations down to 6,000 m. *Nautile* weighs 18.5 tons but can be deployed from a support ship with a relatively small payload.

The manned submersible *Nautile* (IFREMER)

An archaeological survey involves the documentation of as much of the visible site as possible using cameras, drawings, and measurements. A reference grid is typically installed on the site to assist during the registration of the necessary data. The most conventional technique is to place a grid of nylon cables or PVC tubing (2 by 2 or 4 by 4 m) on the seabed. The next step is the methodical exploration of the shipwreck, numbering objects and taking measurements. After that, the objects may be lifted.

In deep water this work may be cumbersome, and a new approach was used on the *Arles IV* site in 1993, called virtual site excavation. This approach made it possible to reconstruct the cargo of this Roman ship by computer. A virtual excavation has several advantages: operations can be conducted at great depths, sites can be studied without raising or displacing objects, and the complete site record is stored in the computer, allowing archaeologists to work on it as many times as they wish.

To complete a virtual excavation it is necessary to study the different layers of the wreck site using an elaborate photographic process called stereo photogrammetry. This process is often used to make geographic maps or drawings of industrial installations. The technique borrows from a basic principle of human vision. Any object that is photographed from two viewpoints and viewed with both eyes through an appropriate optical apparatus appears to be in relief. It is possible to use this visual capability to convert the relief image into three-dimensional data (shape, position, and dimension of the object). The information can then be processed by computer to produce an exact digital model (Blot, 1996).

Before the *Arles IV* site was recorded with a stereo camera mounted on the *Nautile*, graduated rulers and floaters were placed on the site as a dimensional reference grid.

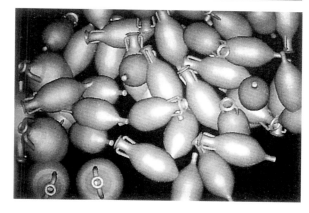

Virtual excavation. The *Arles IV* wreck was first photographed with a stereo camera (top), and the amphora mound was later recreated digitally (middle and bottom) (DRASSM/IFREMER)

Nautile then took three series of photographs of the site, flying at a constant altitude of 3 m above the site and taking a new photo every meter. Each photograph overlapped the next by at least 60 percent, and nearly one hundred photographs were made. An LBL underwater positioning system was used to establish the exact location of *Nautile* and the images with decimeter accuracy.

A sophisticated stereoscope was later used to view each pair of images in relief. Linking this stereoscope to a computer made it possible to measure a photographed artifact in three dimensions and to find its relative position. The ten types of amphoras found on the *Arles IV* wreck site were reproduced in the computer and stored in a computer library. After determining the type of amphora and the exact spatial coordinates from the photogrammetry system, the computer copy from the computer library could be placed in the correct location in the computer model. Doing this amphora by amphora, a three-dimensional model of the *Arles IV* site could be constructed. This virtual excavation is somewhat similar to the photomosaic models created by other projects. The main difference is that by using stereo photographs it is also possible to obtain exact measurements from the model.

Roman Wreck at Cap Bénat

This site was discovered in 1977 by the French Navy submersible *Griffon* at a depth of 328 m. The site consists of a mound of Italian wine amphoras. The first investigation was carried out by an archaeologist from DRASSM who accompanied the Navy on the *Griffon*. DRASSM examined the wreck again in 1980, this time using the manned submersible *Cyana*. Together the two investigations, which each lasted only a few hours, allowed the dimensions of the site to be roughly assessed and photo and video recordings to be made. Fourteen amphoras (of a total of about 350) were removed by a manipulator arm and placed in a geologist's net that had been dragged close to the wreck site by the support ship; the net was then hoisted to the surface, where the amphoras could be checked for stamps or other marks (Blot, 1996).

The *Sainte-Dorothéa*

The *Sainte-Dorothéa* was a Danish merchant ship that sank in 1693 at a depth of 70 m in the bay of Villefranche. The ship was loaded in Spain before it ran aground and was en route to Genoa, after a port of call in Marseilles (L'hour, 1993). It was investigated with the manned submersible *Remora*.

Remora is a single-place manned submersible developed by COMEX. It can reach depths down to 350 m and can stay submerged at a site for hours to conduct observations. The *Remora* has a habitable hull with an interior diameter of 1.2 m, consisting of a transparent acrylic dome in front (70 mm thick, tested to depths of 1,000 m), and a stainless steel dome in the back. The crescent-shaped porthole offers excellent visibility.

Propulsion is by two 2CV motors, and it has an active rudder motor with manual and autopilot modes. *Remora* is permanently attached to the support ship with a 450 m umbilical cord through which surface-generated energy, communications, and control, video, and scanning sonar signals are transmitted. The atmospheric control in the submarine is achieved by a traditional oxygen supply (ca 100 hours), emergency bottles, soda lime blowers for absorption of carbon dioxide (energy and granule capacity, 100 hours), and a dehumidifier to absorb condensation. The pilot can also detach the umbilical cord, using remotely controlled pliers cut the electrical cables where they enter the hull.

Remora allowed the archaeologists to work on the wreck site of *Sainte-Dorothéa* for several hours at a time and to conduct several general and specific observations. The wreck is tilted toward starboard. It seems to be in excellent condition, probably because it is mostly covered by silt. Some of the cannon of the starboard battery are still intact, their muzzles resting on the planking and the opposing side, and several other pieces of artillery have dropped slightly astern.

The *Sud-Caveux*

A new generation of the *Remora* has also been developed. *Remora 2000* is autonomous and can take two passengers to a depth of 610 m. It was used to investigate a Roman wreck/amphora carrier, christened *Sud-Caveux*, outside Marseille at 64 m. The site is about 15 m long. DRASSM used this site to develop new methods for investigation of sites in deep water in cooperation with COMEX (Croizeau, 1996).

The site was first thoroughly documented using the virtual excavation technique developed on the *Arles IV* site. A metal cage was placed on the seabed as a reference point, and 70 mm black-and-white cameras mounted on the *Remora 2000* were used to take stereo photographs. These were later used to create a computer model of the site.

The *Remora 2000* manned submersible (COMEX)

When this phase was completed, an attempt was made to excavate the site. Propellers (15 kW) were lowered down to the site and used to blow away the silt. The top layer of soft mud was removed easily, but the next layer was much harder. It was, however, also blown away when the power was increased. According to the archaeologists, nothing was damaged during this operation. Artifacts that appeared as the mud was blown away were raised to the surface by divers, since this proved quicker than using the *Remora*. As with all the ancient wrecks examined by the French in deep water, all organic material such as wood had disappeared as a result of wood borers and other site formation processes. A computer program catalogued the objects, with pictures, dimensions, positions, and where the items were conserved and stored. DRASSM's ROV, *Comex-Pro*, was also used during this operation, but mainly as an observation vehicle (Croizeau, 1996).

The Basses de Can Shipwreck

This Roman wreck at Basses de Can, south of the Cap de Saint-Tropez, is resting in an area measuring 80 by 90 m. It was assessed in 1987 with the single-seat manned submarine *Nérée*. Archaeologists directed the operations from the surface, via a video screen and intercom. Several amphoras were recovered with a hinged fork tool fixed to the front of the small submarine.

To document the site, 2 m long graduated metal rulers were placed on the site by a remotely operated arm on the submarine. Three reference grids also served as a basis for the map and marked the longitudinal axis of the site. Along this alignment, the rulers were positioned, one by one, with the manipulator arm, running along the whole length of the site. While the pilot later simultaneously manipulated the heading and trim of the *Nérée,* video cameras recorded the site and the rulers. Good measurements could then be obtained from the video recordings and used to produce a site plan.

The Batéguier Shipwreck

This ship was originally from Andalusia. It is found in the Bay of Cannes at a depth of 58 m. In 1973 and 1974 it was the object of salvage attempts, and essential parts of the cargo such as grindstones, metal cauldrons, lamps, and terra-cotta dishes transported in large crates were raised. In 1993 the remains of the wreck, about 20 m long, 4.5 m wide, and 40 cm high, were investigated using a ROV system. The sediment covering the wreck was cleared away using a thruster, similar to the approach used at the *Sud-Caveux* site.

CSS *Alabama*

The Confederate State Ship *Alabama* was built in 1862 at Birkenhead, by John Laird and Sons shipyards. During the Civil War, the *Alabama* sank sixty-four merchant ships and one Union warship. Before long it was sunk itself, off Cherbourg, during an altercation with the USS *Kearsarge.* The shipwreck was discovered in 1985 and is the property of the United States. A French-U.S. team has been undertaking excavation work on it since 1988.

The CSS *Alabama* lies in an area of extreme environmental conditions. Strong alternating currents can reach a velocity of 4 knots, the depth is 60 m, and the visibility is poor. This has made the investigation precarious and demanded the use of advanced methods. The team has used a remotely controlled, battery-powered, bottom-crawling robot with belt drive, the Lagune, in the investigations. The Lagune is capable of moving around on the seabed; navigating and positioning; exploring, observing, and recording video images; picking up objects; and carrying and using a system for ejector excavation.

A lengthy period of evaluation yielded a general map of the site, and a subsequent excavation revealed several artifacts that were subsequently raised to the surface, including a pivoting Blakely cannon, several objects from the officers' mess, as well as personal effects such as fashion accessories, tools, revolver bullets, a sperm whale tooth, and coins. This work was accomplished using a combination of divers and the Lagune.

The "Grand Ribaud F" Etruscan Deepwater Wreck

DRASSM and CNRS are continuing to develop their "digital excavation techniques" as used at the *Arles IV* site. This work focuses on developing a complete shipwreck data management system, based on three-dimensional visualization systems, three-dimensional measuring tools, and object modeling (Drap and Long, 2002). In 1999, COMEX discovered a wreck loaded with Etruscan amphoras and a general cargo situated in more than 60 m of water off the island of Grand Ribaud. Using the research vessel *Archéonaute* and Comex dive support vessel *Minibex*, with its submersible *Remora 2000* and Super Achille ROV, a prop washer named Blaster was used to uncover the sediment. Photogrammetry was then used to record the complete wreck, and a three-dimensional computer model and data management system to keep track of the visible artifacts were later established.

The Lipari Shipwreck

The Institute of Nautical Archaeology (INA) at Texas A&M University has been instrumental in the development of marine archaeology but has mainly been involved in projects within the diving range. Several shipwrecks investigated by INA do, however, rest in relatively deep water. In 1976 and 1977 archaeologists from the Institute partially excavated an ancient shipwreck near Sicily in partnership with Sub Sea Oil Services saturation diver-trainees. The shipwreck was discovered in the mid-1960s and partially looted through 1971, with periodic confiscations of illicit material by the Italian government. The shipwreck lies in 60 m of water on a steep slope 300 m off the southeast coast of the island of Lipari. The ship dates to the third century BC on the basis of several hundred examples of Campanian A black-glazed pottery and wine amphoras (Frey et al., 1978).

During the 1976 and 1977 field seasons, INA archaeologists supervised the excavation of the shipwreck by ten saturation diver trainees working from a diving bell. The archaeologists accessed the site by means of voice communication and underwater video cameras and were able to view the site directly from both a pressurized diving bell and a PC15 manned submersible. Divers collected detailed orthogonal photographs of the shipwreck site, including significant portions of the ship's hull uncovered during the excavation, and video imagery was collected by the PC15 that was essential to drafting a site plan.

Fifty-two intact Greco-Italic amphoras, most likely produced in Sicily or southern Italy and carrying wine, were removed from the wreck site. Also recovered from two deposits were 107 intact or nearly intact bowls and miscellaneous other Campanian and pre-Campanian black-glazed ware, some still neatly stacked. Six square meters of hull was also uncovered, along with additional amphoras and Campanian pottery. The strakes of the hull were joined by pegged mortise-and-tenon fastenings. Frames were attached by both iron and copper nails. Several fragments of lead anchor stocks, as well as an iron anchor, were also found. Analysis of the site and archaeological material suggest that the ship was approximately 20 m in length or more. The origin and destination of the ship and its cargo remain a mystery.

This is one of the earliest examples of deepwater archaeology, and it demonstrated the archaeologists' need for direct access to deepwater shipwreck sites and underscored the need to "prepare divers archaeologically" for the task at hand, a need that persists today with the application of ROVs and their pilots.

The Jason Project

R. D. Ballard, famous for locating the *Titanic*, was the director of the Center for Marine Exploration at WHOI when he conceived and executed the Jason Project. The project was named after the most advanced and versatile of WHOI Deep Submergence Laboratory's unmanned underwater vehicles. In this combination of WHOI skills and experience in marine science and underwater technology with the communications expertise of Electronic Data Systems and the programming and broadcasting skills of the National Geographic Society and Turner Broadcasting System, schoolchildren were invited to join a scientific expedition at sea while doing projects related to marine archaeology, biology, and geology.

WHOI's ROV *Jason* was designed to work in conjunction with a large towed support vehicle, *Medea. Medea* was also used as a stand-alone vehicle; towed behind a research vessel, it could search for sites with its cameras and sonar systems. *Medea* was used mainly for wide-area surveys and *Jason* for precise, multisensory imaging and sampling. When *Medea* worked in conjunction with *Jason*, it provided sufficient weight to tauten the long fiber optic cable needed for deepwater survey. *Medea* hung vertically below the dynamically positioned research vessel and maintained a watch circle of 12–20 m some meters above *Jason*. A neutrally buoyant tether connected *Jason* to *Medea*, and this decoupled the movement of the long cable. *Jason* could therefore work undisturbed by the cable, while the signals to and from *Jason* were sent via *Medea*'s umbilical to the research vessel. *Medea* was also used to provide high-altitude views of *Jason* on the seabed and carried lights to complement those on *Jason*. Both *Medea* and the 1,200 kg *Jason* could work in depths down to 6,000 m (Ballard, 1993).

A special underwater positioning system, called EXACT, was used to position *Jason* and *Medea* relative to a network of transponders placed on the seafloor. In addition to the x,y,z coordinates from the EXACT system, *Jason*'s control sensors included instruments that measured acceleration and attitude. Heading was determined by a flux-gate compass and a directional gyro. The acceleration was measured in three axes with a servo accelerometer, and pitch and roll were measured with a two-axis inclinometer. Depth and the vehicles altitude off the bottom were also recorded. Using special software developed by WHOI, the ROV operators could use these inputs together with GPS positions of the research vessel to show where *Jason* was and had been and where video, photos, and objects were taken and sampled.

Using this positioning input, *Jason* could be programmed to follow precise tracklines automatically to survey wide areas of the seabed, or it could be used in joystick mode to maneuver precisely in small areas such as a wreck site using its seven thrusters. High-sensitivity color video cameras (one of broadcast quality) and a 35 mm still camera were mounted on *Jason*. Black-and-white footage could be shot from the towed support vehicle with an SIT camera.

Jason's first archaeology project was carried out on a wreck in the Mediterranean in 1989. The schoolchildren prepared by studying a curriculum developed by the National Science Teachers Association and then visited fourteen museums where they watched a live broadcast of activities both in the control room of the research vessel and on the seabed. The students were allowed to ask questions, and some were even given a chance to operate the ROV from the museum in the United States thanks to fiber optics, computers, and satellites. Nearly a quarter million schoolchildren attended this demonstration of marine archaeology. The Jason Project has since been made an independent not-for-profit organization—the Jason Foundation for Education—which has continued to create similar projects.

This archaeological site had been identified by Ballard in the spring of 1988 by the towed survey system Argo, equipped with a digital electronic still camera and black-and-white SIT video cameras. The site was located in international waters off the northwest tip of Sicily. A long reef known as Skerki Bank lies about 25 nautical miles south of the site and has been responsible for many ancient and modern shipwrecks. The area showed an accumulation of archaeological material in depths between 750 and 800 m, identified on the basis of scattered ancient amphoras and the remains of a separate ancient shipwreck nicknamed *Isis*. Some eighty-four live programs were broadcast during the fourteen-day expedition at this site. The first week was devoted to the live television broadcasts and the second week to survey work.

The shipwreck *Isis* was the main target of the live broadcasts. The site consists of a group of largely complete amphoras and other archaeological material. The visible surface material of the wreck covers about 10 by 10 m. It was mapped and photographed by *Jason*, and a large mosaic image of the site was later made from the photographs. Forty-eight of the artifacts were chosen for lifting, including ten mostly complete amphoras, five complete common wares, the upper part of a cylindrical millstone, a lamp, sections of iron anchors, and wooden planking. On the basis of this investigation it is believed that the ship had a length of 12–15 m and is from the fourth century (McCann and Freed, 1994). Even though several marine animals were observed on the wreck site, no wood-boring teredo worms (*Teredinidae*) were identified in the recovered wooden planks as expected. However, a rather large wood-boring pholad of the genus *Xyloredo* was found and is probably the reason, with few exceptions, organic materials are absent at the surface of the site.

After this investigation of *Isis*, a 6 by 5 km area surrounding the site was surveyed by *Jason*. The ROV searched systematically in a grid pattern, moving at about 0.5 knots on lines spaced 25 m apart using sides-can sonar, scanning sonar, and cameras. This survey revealed some seventy amphoras, which were mapped and documented with video and still photography. Seventeen of the amphoras were removed for further study. The amphoras were scattered all over the area but mainly in a line running northwest-southeast, which may reflect an ancient trade route from Carthage. Within the area six possible wreck sites were identified, using a criterion of three or more amphoras of the same or similar type found together (McCann and Freed, 1994).

In the 1989 season no digging or imaging beneath the surface took place and the project was limited to survey and recovery of selected archaeological materials for identification and study. To recover the archaeological material, a manipulator arm capable of lifting up to 100 kg was used. Two special end-effecter assemblies were designed to lift the Roman artifacts, cradling them gently in soft, synthetic fish netting. A specially designed autonomous elevator device was used to bring the materials to the surface. The elevator was made of aluminum, with four netted compartments and a glass float. The elevator would free-fall to the bottom while its location was monitored by the underwater positioning system. Operators could not control the descent, but by carefully calculating currents they typically landed the elevator about 50 m from the site without causing damage. *Jason* and *Medea* could use the elevator by moving the research vessel to either pick up or store items. The elevator was later ordered acoustically to rise to the surface by releasing a weight and was retrieved on the surface and hoisted on board the research vessel. This elevator worked well, and forty-eight objects were recovered without damage.

In 1997 the Jason Project revisited the *Isis* wreck to investigate it and surrounding areas even better (Ballard, 1998; McCann, 2000, 2001; McCann and Oleson, 2004). *Jason* and *Medea* were operated from the research vessel *Carolyn Chouest* and used to investigate several archaeological sites that had been discovered by the U.S. Navy's research submarine NR-1 in 1995. The advanced control system of *Jason* allowed the ROV to move along programmed tracks at a constant altitude above these sites and take photographs with a vertical-pointed digital camera. *Jason* was positioned by an EXACT positioning system (50 m range) and a second, long baseline (LBL) system. Keeping a constant altitude above the sites and with a known coverage of the still camera, a computer program could draw the coverage of each photo on top of a site plan, to ensure that the photo survey had enough overlap to construct a complete mosaic. This program

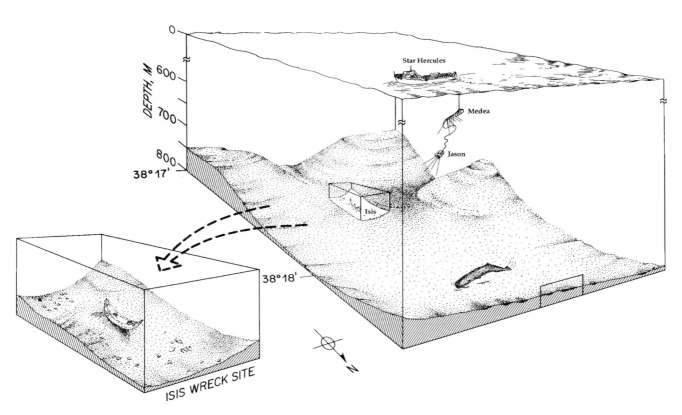

Map of the *Isis* wreck site (from McCann and Freed, 1994)

Jason over the *Isis* wreck site (WHOI)

stored information about each photo (where it was taken, the position, etc.) and drastically decreased the time it took to complete a mosaic. The mosaics were made in a software package developed by WHOI and used to create excellent photomosaics of the *Isis* site. These mosaics are fairly consistent with respect to scale, but exact measurements cannot be made from them because of some height differences at the site. An Imagenex scanning sonar system mounted on *Jason* in a down-looking mode enabled a microbathymetry image to be made of the sites (Ballard, McCann et al., 2000; Mindell et al., 2004).

The U.S. Navy's deep-diving nuclear research submarine NR-1 was also part of the equipment package used in 1997. When a new site was discovered by NR-1, *Jason* was sent down to inspect it. A total of eight wrecks were documented: five Roman sites (including the sites found in previous seasons), a seventeenth- or eighteenth-century shipwreck, and two shipwrecks from the nineteenth or twentieth century, all found within a 100 km² area.

The seafloor in the area is composed of fine grained sediment, and an attempt was also made to excavate one site using a 10 hp dredge mounted on NR-1. This operation was not successful, for the dredge could not be controlled as needed to excavate the delicate objects. The dredge simply made a large crater in the seabed and would probably have damaged the site if the attempt had not been abandoned. Instead, a shovel was attached to the manipulator arm and moved from side to side to brush away the sediments. This worked well on the uppermost layers but was unable to penetrate very deep. The *Jason* team was therefore not able to determine if organic material had survived beneath the surface. Eventually, 113 artifacts were recovered from the surface of these sites and later conserved. The Isis project was probably the first real deepwater archaeology investigation, with its advanced documentation and recovery of artifacts.

Project Archeomar

In 2004 the Italian Ministry of Cultural Heritage invited tenders from throughout the world for a contract to record elements of underwater archaeological heritage in the southern Italian regions of Campania, Basilicata, Puglia, and Calabria. The project was won by a group of eight companies each specializing in the various sectors of archaeology, information technology, and maritime diving and investigation.

The survey at sea was principally conducted by three scientific research vessels (*Minibex*, *Janus*, and *Coopernaut Franca*), equipped with modern and sophisticated instrumentation used for maritime geophysics: sidescan sonar, multibeam, sub-bottom profiler, and magnetometry. For the visual documentation of the sites, three ROVs of the Super Achille class were used, together with a two-man submarine, *Remora 2000*. The surface positioning utilized GPS, and the positioning of underwater vehicles and instruments (submarine, ROV, towfish) was obtained through the use of SSBL underwater acoustic positioning systems. With this instrumentation, it was possible to record video and photographic footage. On board, apart from the crew, were underwater archaeologists, geophysicists, geologists, technicians, and ROV operators.

For each site located, an archive of information was created, consisting of figures, photographs, films, documentary records, and any other type of information considered important and relevant to the site and to the project. The GIS of the Archeomar project was created using ESRI software. The conceptual design of the architecture of the GIS is based on a client/server model, which has permitted, among other things, great flexibility for the insertion of data, the updating of records, and the distribution of information. The system can be accessed through WebGIS.

The complete graphic and photographic documentation now consists of nearly two thousand photographs, nine hundred films, eighty sidescan sonar images, ten sub-bottom profiler images, and four photographic mosaics. In total, 287 underwater archaeological sites were located and added to the database, and a further 476 have been inserted for which there is bibliographic and archival information.

Multiple Ancient Shipwreck Site off Gozo, Malta

A systematic underwater archaeological survey of the Maltese archipelago was initiated in 1999 (Søreide and Atauz, 2002; Atauz, 2008). The sea around these islands gets deep very fast. In addition, looting of underwater sites here has been a serious problem. Therefore, after several years of investigation in Malta with no new shipwrecks discovered in shallow water, it became obvious that possible shipwrecks would be situated in relatively deep water off the coast and out of reach of conventional divers. The INA, NTNU, ProMare, and several

The entrance to Xlendi (Fredrik Søreide)

local institutions therefore joined to survey the seabed in selected deepwater areas.

This effort immediately yielded interesting results. An underwater site best described as a "scatter of several thousand amphoras" was located and surveyed at a depth of 100–120 m outside the ancient harbor of Xlendi. The site covers an area of 100 by 400 m. The majority of the amphoras date from the third century BC and provide the first archaeological evidence of ongoing trade in Maltese waters during the Punic Wars between Rome and Carthage.

Punic settlements indicate that Xlendi was the only sheltered anchorage on Gozo and along the western Maltese archipelago in antiquity. The multiplicity of amphoras and other ceramic artifacts at the site represent ships headed for the bay when they were caught in storms or otherwise wrecked; it should be noted that the estimated size of the contemporary Gozitan population makes the island an unlikely final destination (Peacock and Williams, 1986).

The Maltese archipelago was first colonized in the sixth millennium BC by settlers who reached the islands on an unknown type of watercraft, most likely made in Sicily. Successive waves of colonizers, attackers, immigrants, and countless travelers who only touched at the islands similarly arrived by sea. The development of underwater archaeology in Malta is of crucial importance not only because the archipelago received all of its occupants and cultural influences by sea but also because underwater material might be the only available archaeological record.

Malta is a good case study to illustrate the application of deepwater archaeological techniques. Wind patterns around the archipelago are unpredictable, to say the least. The strongest and most hazardous winds are from the east and northeast, and to avoid the dangers presented by sudden storms seafarers were at times forced to sail along the southwestern coasts of Malta and Gozo. This coast is lined with high cliff faces towering to more than 250 m above the sea, and it does not provide any safe anchorages. Therefore, this portion of the coastline had a high potential for shipwrecks. At the same time, though, this coastline gets very deep very fast. The cliff face continues underwater and reaches 80–100 m at 50–100 m off the coast.

One of the most promising areas selected for the survey of 2001 along this coast was the entrance to Xlendi Bay, on the island of Gozo. This inlet is the best-protected anchorage for the island of Gozo, and the discovery of Punic tombs near the inlet supports the archaeological

Xlendi: The ROV on the deck of the survey vessel (top), and team observing the video feed from the ROV below (bottom) (Fredrik Søreide)

The large amphora mound found off Xlendi, Malta

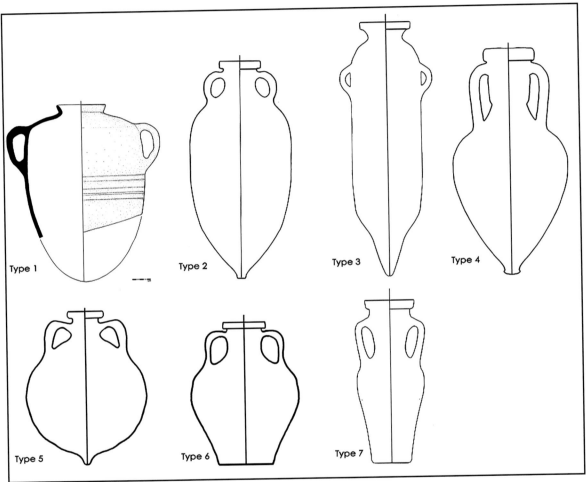

Main amphora types represented at the Xlendi site (Ayse D. Atauz)

potential of the bay. The cliffs on the coast allow entrance to the small inlet with a small beach, but there are two reefs at the entrance that are hazardous for ships. The survey began from the outer reef and covered an area to the maximum depth of 200 m.

The initial survey relied on scanning sonar that was part of the ROV equipment; ROV cameras immediately inspected anomalies detected by the sonar. The survey was carried out using an electric ROV with sufficient flexibility, power, and size to complete advanced documentation tasks and even manipulative tasks required for sampling in fairly deep water. In addition to the scanning sonar and video cameras, the ROV was equipped with an underwater positioning system that defines the positions of the ROV and objects located relative to the survey vessel. The equipment also included a laser-based underwater measurement system, which was used to measure artifacts found at the site.

The major find of the 2001 season employing this technology was an amphora scatter off the entrance to the bay—several thousands of amphoras representing at least seven different types. The amphora field is spread over an area of about 100 by 400 m. The scatter is in the middle of flat, sandy bottom at a depth of 110–130 m and about 6 km off closest landfall. It does not continue toward land, although fragments and complete examples of amphora types represented at the deep site were recovered from the shallow reef of Xlendi in the 1960s by British Navy divers.

The variety of dates represented by the amphora types found in the Xlendi site suggests several shipwrecks (Caravale and Toffoletti, 1977; Ayuso and Bernal, 1992). Only one particular ovoid Punic amphora was recovered; this dates to the third century BC and is likely the product of a workshop in western Sicily or the vicinity of Carthage. This amphora type has a wide distribution pattern in the Mediterranean and was found in sites from the Atlantic coast of Spain, the Balearic Islands, near Carthage, and also in Punic tombs in Malta. The second most common amphora type at the Xlendi site also dates to the late third or early second century BC. It is likely that this Punic type was produced in Tripolitania or western Sicily and is found in archaeological contexts in Spain, the Balearic Islands, Sardinia, Corsica, and sites on the southern coast of modern France and the Italian peninsula as well as in Tunisia.

The other types on the site are more problematic in terms of dating, but it seems plausible that some types are more recent than those mentioned above. Two of these are likely to date to the late Roman period of Malta, between the first and third centuries. Overall, amphoras on the site range from the fifth century BC to the fifth century AD, a timespan of 1,000 years, indicating that the scatter consists of several shipwrecks, at least one of them dating to the third century BC.

As everywhere else in the Mediterranean, the third century BC was a period of crisis brought about by the First Punic War. Punic Malta became involved in the territorial scenario of the war because of its strategic position at the middle of the Mediterranean, forming a perfect Carthaginian base for attacks on Sicily and southern Italy.

The Maltese islands were captured during the Second Punic War (218 BC) by a Roman fleet that disembarked from Lilybaeum in Sicily. The Romans had two motives in invading Malta. One was to prevent its use as a Carthaginian naval base. The second had deeper roots: Malta was used as a pirate base in this period, where the pirate ships wintered and sought shelter when in distress. These pirates not only harassed Roman commercial ships and threatened the security of navigation close to Italian and Sicilian coasts but also threatened supply lines to North Africa, of crucial importance during the war. The insecurity and instability of this period are detectable through the archaeological record in the construction of defensive walls and watchtowers. A detailed study of the Xlendi site is therefore crucial to an understanding of Malta's role within this struggle between Rome and Carthage.

This survey showed that the potential for deepwater research in Malta is high. The unique multiple shipwreck site is the perfect proving ground for subsea mapping and excavation technologies and could become a watershed for ongoing work in Malta and other submerged sites.

Eastern Mediterranean Sea

In recent years several deepwater archaeology projects have also been carried out in the Eastern Mediterranean.

Phoenician Shipwrecks off Israel

In 1998, R. D. Ballard left WHOI and established the Institute for Exploration at the Mystic (Connecticut) Aquarium. In June 1999, the Institute mounted an expedition to the eastern Mediterranean Sea. The primary goals of the archaeological research in 1999 were to survey, plan, and photograph two shipwrecks and to collect samples of artifacts and other relevant material from their cargoes to learn more about the wrecks (Ballard et al., 2002).

The two shipwrecks were first discovered west of Israel by the U.S. Navy's research submarine NR-1 in 1997. Both ships appear to be of Phoenician origin from the eight century BC, the earliest known shipwrecks to be found in the deep sea. The ships were probably laden with cargoes of fine wine destined for either Egypt or Carthage, when they were lost in a storm on the high seas. The ships lie upright on the seafloor at a depth of 400 m in a depression formed by the scour of bottom currents. Their discovery suggests that ancient mariners took direct routes to their

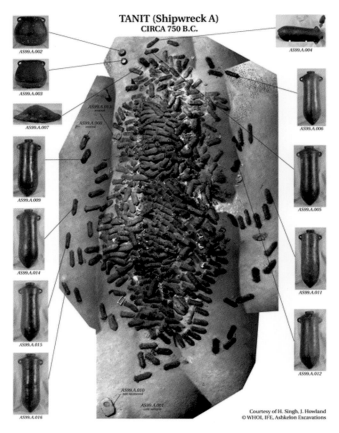

Photomosaic from the Phoenician wreck site off Israel (Hanumant Singh, WHOI)

destinations even if it meant traveling beyond sight of land.

The team used WHOI's *Medea/Jason* remotely operated vehicle system to investigate the sites. A 120 kHz phased array DSL-120 sidescan sonar system was first used to relocate the shipwrecks on the seafloor using sonar and multibeam. When the sites were relocated, *Jason* was sent to the seafloor to inspect it. The ROV placed an array of EXACT transponders around the site, and this enabled the vehicle to operate in closed-loop control and conduct a series of survey lines 1–2 m apart. The navigation suite used consisted of the bottom-installed EXACT transponders, an attitude reference package, a precision depth sensor, and a 1200 kHz bottom-track Doppler navigation sonar system mounted on *Jason*. Each survey line was used to collect images for a photomosaic while 675 kHz digital sonar was scanned back and forth to measure the micro-topography (Foley and Mindell, 2002).

The optically derived photomosaics provide a vivid view of the shipwreck. Because of incremental errors in building up the photomosaic, one cannot make precise measurements over the entire mosaic. The bathymetric map, on the other hand, though not as visually appealing provides a mechanism for making precise quantitative measurements across the entire site.

After completing this survey, the archaeological team selected specific artifacts for recovery on the basis of diagnostic qualities to help determine the age, origin, and significance of each wreck. To avoid damage to the site, only artifacts lying along the perimeter of the site were initially selected. The main recovery tool was a simple hydraulic device that consisted of two opposing horizontal tongs each having a webbed mesh net. With the tongs in a relaxed position, the operator slid the lower tong beneath the desired object, closed the upper opposing tong using the hydraulic actuator, and then thrust the vehicle up off the bottom until it was free of the site. The pilot then drove to an elevator and placed the artifact in a selected and numbered compartment. Some artifacts were also collected while *Jason* was hovering above the site in closed-loop control. The artifact was then collected by the manipulator arm, carefully adjusting the force not to break the artifact. The most common artifacts were amphoras. The team collected twenty-four amphoras from the two sites in addition to a small number of additional artifacts.

One of the shipwrecks has been named after the ancient goddess Tanit. Four hundred amphoras could be seen on its upper visible surface, and that particular amphora style dates to 700–750 BC. Each amphora carried 18–19 liters of liquid, so the ship carried an estimated 11 tons of liquid, probably wine. The amphoras have small handles, which would have been used to rope them together during transit. A 1.5 m, one-hole stone anchor was also found. This is the first Iron Age anchor to be put in the context of a ship. The ship is an estimated 16 m long and a third that length wide. This shape is documented in models from that time period, which are characterized by the 1:3 ratio of width to length.

The second wreck, which the team named *Elissa*, contained 350 visible amphoras amid chaos of shrimp. This ship is dated to ca 750 BC. There are two anchors amidships on each side of the vessel. The team found galley equipment and ballast stones in the bow end. The ship is estimated to be 18 m long and 6 m wide. The team also found a grinding bowl that was a possible import, cooking pots, a challis and incense stand, and wine decanters that identify the vessel as Phoenician.

In 2003 the team tried to go back to the site for more detailed archaeological investigations but could not obtain a permit from the Egyptian authorities.

The Greek-Norwegian Deep-Water Archaeological Survey

Large areas of the Greek coastline have been examined by divers, and Greek authorities have an existing database of around a thousand shipwrecks in shallow water. Little is known, however, about sites in deeper water where conventional diving is impossible. The Greek-Norwegian Deep-Water Archaeological Survey is a joint project of the Greek Ephorate of Underwater Antiquities (EEA), NTNU, and the Norwegian Institute at Athens (NIA). The main goal of the project was to develop and field-test a set of practical and cost-effective remote sensing methods that can be used by most archaeological institutions. From 1999 to 2003 the project surveyed the seabed in selected areas around Greece to locate marine cultural remains in depths below 50 m that had not been explored by divers. The results of this survey constitute a new source for research activity and an effective tool for the Greek Ministry of Culture in its efforts to protect the underwater cultural heritage (Delaporta et al., 2006).

Survey Methods This project was the first systematic use of underwater survey technology in Greek waters. One of the main goals of the project is to keep the costs to a minimum. Most authorities and archaeological institutions refrain from mapping their cultural heritage in deeper water because of the perceived high cost. Archaeologists often consider deepwater survey technology a fascinating but complex and expensive high-tech tool, and it is seldom implemented in archaeological activities. This project was therefore designed to show that it is possible to perform systematic large-area surveys using existing equipment from small ships of opportunity with a small crew at relatively low cost with good results. By locating and documenting underwater archaeological remains in deeper water, the Greek authorities are building a database of deepwater sites that they can use for sound underwater cultural heritage management. Ideally this will stimulate similar activities in other countries.

Throughout the survey the project utilized a SeaScan PC high-resolution sidescan sonar system from Marine Sonics operating at frequencies of 150, 300, and 600 kHz. This sonar system was towed behind a fishing vessel and used acoustic signals to construct an image of the seafloor and objects lying on top of the seafloor. The lowest of the three frequencies was employed for large-area surveying and was set at range of 100 m to each side of the towfish and towed at a speed averaging 2.5 knots, 10 m above the seafloor. The highest frequency, 600 kHz, was utilized for detailed site imaging of shipwrecks located during the large-area survey and was set at a range of 50 m to each side of the towfish and towed at a speed averaging 1.5 knots, 5 m above the seafloor. The large-area survey covered nearly 200 km of survey lines.

The SeaScan PC sidescan sonar was operated by two project personnel for a total of six weeks during the three seasons that field research was undertaken. Data were collected on a portable computer and coupled with differential GPS. All targets were identified real-time and exported to a simple database program to manage the several hundred anomalies detected throughout the course of the survey. In the evenings the sonar data were reviewed by the sonar operators and the database was modified to highlight the most promising targets for further inspection at the end of the large-area survey.

For visual inspection of chosen targets the project utilized ROVs from Sperre and SeaEye. Two of the most promising targets were verified as shipwrecks by divers in the North Sporades area, and one shipwreck site was verified by ROV in the Ithaki area. An additional seventeen targets were inspected by ROV but found to be geological features. The ROV inspection was carried out to confirm targets, and only brief video documentation was carried out at each target. The most promising targets may be selected for more detailed documentation at a later stage.

Field Activities in the Northern Sporades Area Field activity was inaugurated in 1999 in the Northern Sporades area. The need for raw materials and marine resources had turned the first inhabitants of the Aegean toward the sea. Earliest evidence of seafaring comes from the Franchthi cave in the Argolid (Peloponnesos), where obsidian from Melos was found in strata dated around 11,000 BP. This valuable volcanic rock was distributed from Crete to Macedonia, implying deliberate sea voyages at that early date.

No evidence of such early seafaring has ever been discovered. During the Greek Classical period (fifth to fourth century BC), trade was dominated by the Athenian

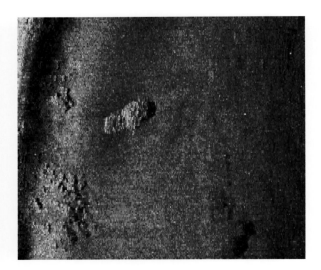

Sonar image of the Byzantine shipwreck discovered in Vasiliko Bay (Fredrik Søreide, NTNU)

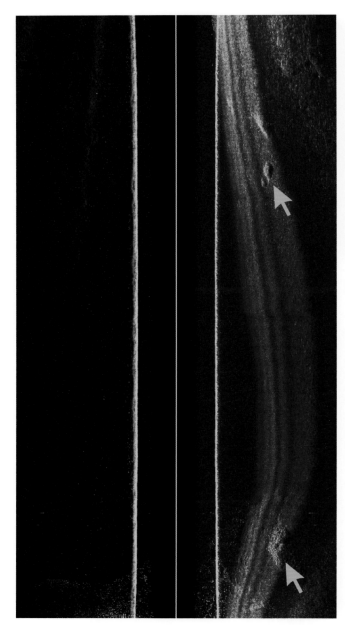

The "Alonnesos wreck" (bottom arrow) and the remains of a small Byzantine shipwreck (top arrow) (Fredrik Søreide, NTNU)

fleet, which ranged from Massalia to Byzantium and the Black Sea carrying mainly grain and wine in amphoras. Amphoras from Mendi (Chalkidiki in northern Greece), Chios, and Peparethos, together with fine Athenian ware, are the main cargo of numerous wrecks recorded by the EEA around these islands, showing that they were part of the Athenian trade network. The most important of these wrecks is that from Peristera dated to the late fourth century BC, investigated by the EEA since 1992 under the direction of Elpida Hadjidaki, since it offers new information about ship construction and the economy of that era. The cargo of approximately four thousand amphoras (weighing 28–37 kg each) required a merchant vessel capable of carrying more than 100 tons of burden (Hadjidaki, 1996). It seems that such vessels were not uncommon from the fifth century BC onward.

Equally important is the wreck investigated near the islet of Phagrou (or Pelerissa) in the bay of Agios Petros at the island of Kyra-Panagia, which carried a cargo of Mendian amphoras dated to the middle of the fifth century BC, since it is the earliest Classical wreck ever found. The archaeological finds are more abundant from the Early and Middle Byzantine periods (sixth to twelfth century AD).

The survey was primarily carried out using towed sidescan sonar as described above. To familiarize the Norwegian crew with the characteristics of ancient shipwrecks in Greek waters, the sidescan sonar was first used to relocate five known shipwreck sites reported by the EEA. This allowed the Norwegian team to calibrate the sonar equipment to local conditions such as underwater topography and wreck site appearance.

The first target was the shipwreck from the Classical period, which was excavated by Greek authorities under the direction of Dimitris Chaniotis in the 1990s. This is the oldest known Classical shipwreck in the Aegean Sea, dating to ca 450 BC. The site now consists of approximately fifty amphoras visible on the seabed. The other known wreck in this area was a shipwreck from the

Amphora mound of the Byzantine shipwreck in Vasiliko Bay (above) with detail photo of the amphoras (below) (Fredrik Søreide, NTNU)

Middle Byzantine period (ca AD 1050). It was partly excavated in 1970 by the Greek Ministry of Culture under the direction of Charalambos Kritzas and the late Peter Throckmorton. Because only about thirty-five scattered amphoras were visible on the topographically complex seabed in this area, the new project's sonar had difficulties detecting the site.

The project also relocated the large Classical shipwreck (the "Alonnesos wreck") that had been investigated by the Greek authorities under the direction of Hadjidaki. This site consists of a large amphora mound, which was easily detected by the sidescan sonar. Additionally, two large Byzantine shipwreck amphora mounds lying close to each other were found in Vasiliko Bay on the island of Peristera. These two wreck sites were also easily detected by the sonar.

These test surveys proved that the sidescan sonar system could be used to detect the amphora cargoes of the

History of Deepwater Archaeology   47

known ships. On the basis of these attempts it was concluded that the underwater topography, the size of the amphora mound, and the sonar frequency are determining parameters for locating this type of marine archaeological site. With a relatively flat and sandy seabed and large amphora mounds, long-range, medium-resolution (150 kHz) sonar could be used to locate sites and create acceptable images. With difficult underwater topography with rocks and varying depths and low numbers of amphoras visible on the surface of the seabed, higher-resolution/shorter-range 600 kHz sonar was necessary.

After this calibration of the equipment, the crew continued with a limited survey in selected areas off Peristera. This survey concentrated on depths beyond the diving limit, primarily 50 m to about 100 m, in an area previously not surveyed. This work resulted in several new discoveries. Approximately 200 m from the "Alonnesos wreck" the sonar detected a fairly small (about 8 m in length) feature that was consistent with an amphora mound. Diving inspections later confirmed that this is a small Byzantine shipwreck.

Another interesting sonar feature was discovered only a few hundred meters away from the two known Byzantine wrecks in Vasiliko Bay. This target was verified during the survey by Greek divers as a Byzantine shipwreck. The site was found to consist of at least several hundred early twelfth-century amphoras at nearly 60 m depths.

Field Activities in the Ionian Sea   In 2000 and 2003, field activities shifted to the Ionian Sea. The project would make a complete and systematic large-scale survey of the seafloor around Ithaki, in depths beyond the reach of divers, primarily from 40–50 m down to about 100 m. The main survey area was the channel between Ithaki and Kefalonia, which is 2.5 km wide at the northern end and widens southward to 5 km. The average depth is ca 150 m, and 110–120 m at the south end. There are no rocks or shallows except for the islet of Daskalio (incorrectly identified with the Homeric Asteris). These favorable characteristics are somewhat tarnished by irregular currents and gusty winds coming off the high coast.

Ithaki is probably most famous for its connection with Homer's Odysseus, but whether Ithaki was the Homeric Ithaka of Odysseus remains outside hard proof. Nevertheless, the identification has spawned a substantial industry geared toward associating places described by Homer (who had a very vague notion of Ionian geogra-

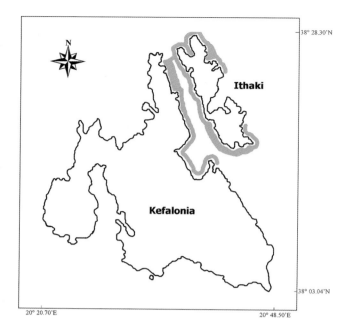

Sonar coverage, Ithaki and Kefalonia (Fredrik Søreide, NTNU)

The ROV is launched to investigate a sonar target, Ionian Sea (Fredrik Søreide, NTNU)

phy) with topographic features on the island. Currently at least two expeditions are hunting for Odysseus' palace, and "the School of Homer" has been identified under the chapel of Agios Athanasios near Exogi.

Other surveys have revealed terrestrial archaeological sites from as early as the Early Helladic. Lying at the rear of the Greek mainland, the Ionian Islands remained outside the main political spheres and economically isolated. Ithaki, or popularly Thiaki, with little change in the name over the centuries (except for the Italian designation of Val di Compare, or Cefalonia Piccola), was in antiquity of minor import—as were the Ionian isles in general—to the cultural development of ancient Greece. Nonetheless it probably was an important station on the ancient route

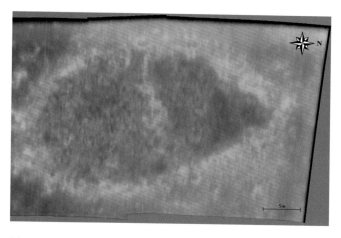 

Photomosaic of Roman wreck, with detail photos of the amphoras (Fredrik Søreide, NTNU)

to Italy. Because of the poor agricultural resources, maritime trade was the prime support for the population, and at the end of the Byzantine era (twelfth century) Ithaki was a notorious pirate lair.

A large-scale sidescan sonar survey was carried out along the coast of Ithaki and Kefalonia to provide a catalogue of acoustic images of the seabed. Several sonar targets were considered potential shipwreck sites. To investigate these further, the team deployed one of the two main ROV systems used during the survey. Capable of operation to several hundred meters, these ROVs were equipped with scanning sonar and a variety of cameras to relocate targets on the seafloor. Additional equipment included a manipulator arm and a laser measurement system that utilizes two lasers positioned in parallel to project a scale of approximately 10 cm onto the seafloor and into the video images. This can also be used to measure distances in the image. GPS was used to position the ship, and a Kongsberg SSBL (super-short baseline) underwater positioning system was used to position the ROV relative to the ship on the surface. The position of the ROV and targets were then fed into a GIS system.

The main discovery was made toward the northern end of the sound between Ithaki and Kefalonia. Based on images from the sidescan sonar and a tip from a fisherman who occasionally caught amphoras in his nets, the ROV was deployed to carry out a systematic survey in the area. A large amphora mound at approximately 60 m was located. The wreck is more than 25 m long, with a cargo of several thousand amphoras. The amphoras indicate that the wreck is Roman and dates from just before to the start of the Christian era. This discovery underpins the assumption that these islands were stepping stones on the important trade route to Italy. The ROV was used to carry out video and photo documentation of the site.

The ROV was also used to document a known shipwreck from the fourth century BC. The ROV carried out a systematic survey of this wreck, which is situated in depths of 20–30 m near the village of Fiscardo on Kefalonia. Several additional sites located by the sidescan sonar have not yet been investigated by ROV.

Overall, the Greek Norwegian deep-water archaeological survey has shown that the potential for archaeological studies in deep Greek waters is high. The project has so far provided a catalogue of acoustic images of the seabed and underwater archaeological sites in the selected survey areas. It has also shown the practicality of performing systematic, large-area surveys using existing equipment from small ships of opportunity, at relatively low cost with good results. Six weeks of survey work covering nearly 200 km of survey lines were completed at a cost of only approximately US$100,000.

Another Ionian Sea Shipwreck

Discovered by ProMare during a geophysical survey aboard the NR-1 in 2001 in the northern Ionian Sea, this shipwreck lies at 800 m midway between Italy and Greece. The morphology of the anchor suggests that the wreck dates no later than the fourth century AD.

The shipwreck is buried in soft sediment and only the edge of the hull is visible on the seafloor. Low-frequency sonar images clearly show that a considerable portion of the ship is present in the sub-bottom and most likely is well preserved. The most recognizable feature on the site is the iron anchor resting on the hull.

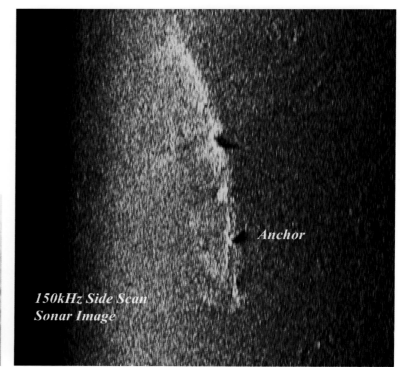
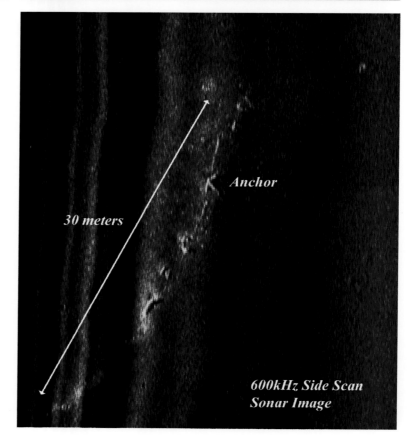

Ionian Sea shipwreck, sonar images (right) and anchor (above) (ProMare)

In Search of Ancient Persian Warships

In 2003 an international research project mounted a deepwater search off the northern coast of Greece in search of a fleet of Persian warships presumed lost in a massive ocean storm in 492 BC. The project was a collaborative effort of the EEA, the Canadian Archaeological Institute at Athens, and the Hellenic Center for Marine Research (HCMR).

The armada is believed to have been sent by Persian king Darius to invade Greece, according to ancient historical accounts. The team's search for the fleet is based on a historical source, Herodotus, whose extensive writings include a report that in 492 BC nearly three hundred ships and more than twenty thousand men perished in a severe storm off Mt. Athos. The event was said to have caused Persian king Xerxes to cut a canal through the narrowest part of Mt. Athos prior to his 480 BC invasion of Greece to avoid the need to round the peninsula in the Aegean Sea.

The team used sonar from the HCMR's RV *Aegaeo*, the manned submersible *Thetis*, and the ROV *Achilles*. The survey area was chosen by the team after two local fishermen raised two Greek bronze helmets in 1999. However, nothing much was discovered except a few amphoras and a bronze point. The point tentatively has been identified as a "sauroter," a bronze spike at the end of a spear. It served as a counterweight and also allowed the shaft to be stuck in the ground when not in use.

WHOI/MIT in the Aegean Sea

In 2005 the EEA and a team from WHOI and MIT surveyed an ancient shipwreck site in the Aegean Sea. Partnered with the HCMR, the team used the WHOI SeaBED AUV to document the wreck site with photography and high-resolution sonar mapping. The team was based on the Greek research vessel *Aegaeo*, operated by HCMR.

The goals of the project were to experiment with and demonstrate advanced technologies and to develop archaeological methods for the AUV. The primary data results included a two-dimensional photomosaic of the shipwreck and a precise bathymetric map of the site.

The shipwreck chosen for this experimental project dates to the fourth century BC. The site is between the islands of Chios and Oinoussia in the eastern Aegean Sea and was discovered during a 2004 sonar survey by HCMR scientists and EEA archaeologists. The wreck is too deep for conventional scuba diving but shallow enough for some ambient light to reach the seafloor. The ship was a merchantman, carrying a mixed cargo probably including wine from Chios and olive oil from Samos. Amphoras are the most visible remains of the shipwreck.

Over two days the AUV performed four missions on the site, repeatedly mapping and photographing the wreck. After completing the survey, the team moved on to inspect the remains of a nineteen-century warship near Chios Town harbor. The HCMR ROV collected video imagery of the wreck's scattered remains, then the team moved to the west coast of Chios, where it used the AUV to survey an early Roman era wreck.

The Chios project showcased the efficiency of AUV survey. In a single three-hour dive, SeaBED's multibeam sonar completely mapped the wreck while the digital camera collected thousands of high-resolution images. Later the same day the team assembled those images into photomosaic strips, giving the archaeologists their first overall views of the wreck. Successive AUV missions on this Classical site provided photographs of the wreck from different angles, showing more details in the artifacts.

The Sea of Crete Project

The Sea of Crete has been a crossroads of maritime activity in the eastern Mediterranean for thousands of years. In 2006, R. D. Ballard initiated the first season of a multiyear geological and archaeological study here, with the goals of describing navigational routes and their hazards; clarifying the nature and organization of communications among inhabitants of the Aegean and Mediterranean seas, including trade and colonization; and correlating the archaeological record with the geological stratigraphy of the Sea of Crete.

In April and May 2006 a team of marine geologists, archaeologists, and engineers boarded RV *Endeavor* to carry out the initial sidescan sonar survey. The team surveyed approximately 500 km^2 and found 203 anomalies. After a preliminary review of the sonar data, the team returned in June to investigate twenty-one targets with IFE's ROV. In total, eight ROV dives were conducted, and the vehicles were working on the seafloor for a total of eighty hours. Three of the targets turned out to be archaeological sites, seven were geological, and the rest either were not located or were modern debris.

One of the first targets investigated, Target 153, was found to be a scattered site of six ceramic vessels and a large concentration of pottery shards. Initial inspection suggests that the ceramics are likely from the Roman or Byzantine periods, but they have yet to be positively identified.

Two modern shipwrecks, probably dating to the late nineteenth or early twentieth century, were also discovered. In addition to target investigation, the team also undertook east-west visual transect surveys with the ROVs in the central Sea of Crete. During these operations they found fifteen or so ceramic vessels that did not seem to be associated with any other archaeological material; most of them date from the Classical to Byzantine periods.

This project will continue the location and identification of archaeological material and further explore the maritime history of the Sea of Crete.

The Navarino Naval Battle Site

The University of Patras has carried out an integrated remote sensing survey in Navarino Bay (modern-day Pylos), where in 1827 a battle was fought between the allied British, French, and Russian navies and the Turkish-Egyptian fleet (Papatheodorou et al., 2005). Integration and interpretation of the remote sensing data showed the presence of shipwreck remains on the seafloor and possible shipwrecks buried under the seabed. The research also showed that these historical remains are under threat from the heavy anchors of tankers that sink into the seabed and, when dragged, dig furrows, thus disintegrating shipwreck remains.

Black Sea

The Black Sea has long intrigued undersea explorers and archaeologists. The region lies at the crossroads of ancient Europe, Asia, and the Middle East. Evidence suggests that it was a center of maritime trade for millennia, stretching as far back as the Bronze Age. The Black Sea is also the only sea with a deepwater anoxic layer, oxygen-deprived waters where wood-boring mollusks cannot survive. In this environment, wooden shipwrecks of antiquity can remain in a state of high preservation for thousands of years.

Northwest Turkish Coast

In 2000, R. D. Ballard's Institute for Exploration conducted a major expedition in the Black Sea along the northwest coast of Turkey from the Bosphorus to the Turkish seaport of Sinop (Ballard et al., 2001; Ward and Ballard, 2004). Research methods included the use of phased-array sidescan sonar, a towed imaging sled, and a small ROV to collect deep-sea survey data. Three shipwrecks and a probable site reflecting human habitation prior to a proposed flooding event were located at depths around 100 m. One additional shipwreck was found within the anoxic layer at a depth of 320 m. The ship found within the anoxic layer was intact, in a state of high preservation, and dated to the Byzantine period around AD 450. The hand-carved mast still soars 10.6 m above the deck, just as it did 1,500 years ago.

In 2003 the team went back to the site (Coleman et al. 2003; Coleman and Ballard, 2004). For this expedition they utilized a new tow sled, *Argus*, and ROV *Hercules*. This system work in tandem. The tow sled hangs on the end of a long cable dangling from the ship and is maneuvered primarily by moving the ship and raising and lowering the cable. Thrusters on *Argus* allow the pilot to aim its lights and cameras toward sites of interest and *Hercules*. *Hercules* is attached to *Argus* via a soft tether and is intended primarily for gathering high-quality video images of underwater artifacts. Pilots can maneuver the vehicles via remotely controlled thrusters. *Hercules* is specially designed for excavation—digging and recovering artifacts from ancient shipwrecks in the deep ocean. It has two mechanical manipulators that can lift, push, grab, and generally do the work involved in excavation. A variety of tools are used, particularly water jets and suction pipes to clear mud away (Coleman et al., 2000).

The main focus was the site dubbed "Shipwreck D." Ballard believes it to be a trader from the fifth and sixth centuries that was trading the Black Sea region. Shipwreck D is so well preserved that cord tied in a V-shape at the top of the vessel's wooden mast is still clearly visible. Also evident are the spars that had rigged the sails as well as bracing timbers and stanchions that still show the wooden treenails that held the ship together. *Hercules* brought up six amphoras and managed to excavate a few minor test pits around the wreck site.

In 2006, Ballard returned to the Black Sea, now working for the archaeological oceanography program at the

Hercules (left) and *Argus* (below) (Institute for Exploration)

University of Rhode Island. The fieldwork was focusing on two areas south of the Crimean Peninsula in the northern Black Sea. The Greek colony of Chersonesos was founded here during the Classical period (ca fifth century BC) and inhabited up through the Middle Ages. The aim of the 2006 survey was to use sidescan sonar and ROVs to locate, identify, and document archaeological sites that might tell us about the patterns of trade connected to Chersonesos, both within the Black Sea and out to the Aegean and Mediterranean. Patterns of ship traffic and cargos from around the Black Sea will help us understand the nature and scope of ancient trade in the Pontus region.

RV *Endeavor* covered approximately 650 km² of the seafloor with sidescan sonar and sub-bottom profil-

History of Deepwater Archaeology 53

The "perfect shipwreck" in the Black Sea (Institute for Exploration)

ing. The team logged a total of 494 targets on the sidescan and also noted several features in the profiles, such as submarine river channels on the Black Sea shelf. Sensors on the *Hercules* collected data during every dive and provided vertical profiles of the oxygen levels across the oxic/anoxic boundary. Dives ranged from shallow depths, just under 100 m, to depths greater than 1,600 m away from the shelf. The wrecks of the warships *Dzherzynsky* and *Lenin* were found and documented with *Hercules*, and other wrecks were also found, including the pre-Dreadnaught warship *Ekaterina*, a medieval amphora wreck, and three aircraft. The medieval wreck was located above the anoxic water layer and is poorly preserved; the jar types have been roughly dated from the ninth to eleventh centuries. The *Dzherzynsky* and *Lenin,* as well as the *Ekaterina,* were found in shallower water, above 150 m. Besides these shipwrecks, other targets turned out to be various bits of trash or cable that show the highly active history of this region. This was some of the first deepwater archaeology conducted in the waters near the Crimean Peninsula and along the trade route between the Bosphorus Strait and Sevastopol/Chersonesos; the medieval wreck lies directly on this line.

R. D. Ballard also returned to the Black Sea in 2007 to revisit Shipwreck D and to characterize the environment in which it rests, to see how the site had changed in the previous four years, and to compare its preservation with that of a Byzantine shipwreck, Chersonesos A, which lies in shallower water off the coast of Ukraine. The project also planned to check out additional anomalies discovered in 2006. Future work will continue to focus on this region in an effort to continue to document this area's maritime history.

Bulgarian Black Sea Coast

In 2002, Ballard's team discovered remains of an ancient trading vessel over 2,300 years old that sank in the Black Sea off the coast off present-day Bulgaria. The vessel dates from the fifth to third century BC. The shipwreck is one of the oldest ever found in the Black Sea. It joins a relatively small handful of other known shipwrecks of the Classical Greek period, including a similar vessel discovered at 60 m farther south in Bulgaria by the Bulgarian Center for Underwater Archaeology in cooperation with NTNU.

Members of the joint U.S.-Bulgarian research expedition discovered the wreck at 84 m off the eastern coast of Bulgaria. Using a three-person submersible vehicle launched from the 180-foot Bulgarian research ship *Akademik,* the team dove on the target previously identified by sonar. The site consists of twenty to thirty amphoras. A large amphora was the sole artifact retrieved. This two-handled clay jar, a type used by Greek and Roman merchants, is unusually large, measuring nearly 1.0 m tall by 0.5 meter wide. Recent analysis of sediment gathered from inside the amphora revealed that it contained bones of a large freshwater catfish species, several olive pits, and resin. Such seemingly small clues have already answered questions about the ship's cargo and possible origin for researchers, while raising others. Cut marks visible on the fish bones, together with other physical clues and references from classical literature, lead researchers to believe that the amphora carried fish steaks, catfish butchered into 6 to 8 cm chunks and perhaps salted and dried for preservation during shipping. The team therefore believe this to be a supply boat full of butchered fish that was being brought from the fish-rich regions of the Black Sea and their associated lakes back to Greece. Additional investigations planned for 2003 were not accepted by the Bulgarian Academy of Sciences.

Mapping the Portuguese Coastline

In 2002, NTNU, ProMare, and CNANS (Portuguese National Center for Nautical and Underwater Archaeology) initiated a large-scale deepwater survey of the Portuguese coastline. The team also decided to survey the seafloor near Cascais in Portugal with the hope of finding the clipper ship *Thermopylae*. After only a brief survey at sea, a large shipwreck was found; divers from CNANS have now confirmed the ship's identity.

Thermopylae was the second composite ship built at Walter Hood's Aberdeen, Scotland, shipyard (1868). The vessel had wooden planking but iron frames. Composite ships were cheaper to build and had a greater capacity for cargo. Cargo capacity would have been quite important, for *Thermopylae* was intended for the tea and wool trades.

Thermopylae soon gained a reputation for speed. On its maiden voyage, the vessel sailed to Melbourne, Shanghai, and Foochow, breaking records on each leg of the journey. The clipper's greatest rival was *Cutty Sark,* but it is uncertain which vessel was faster. The two sailed together from Shanghai in 1872, but *Cutty Sark'*s rudder was carried away, ending the contest.

Despite its fame as a tea clipper, *Thermopylae* more often sailed to Australia in the wool trade. The ship continued to make these voyages until 1890, when it was sold to Canadian owners. In later years, *Thermopylae* was bought by the Portuguese Navy and renamed the *Pedro Nunes*. The vessel was converted to a coal hulk and finally sank in 1907.

Baltic Sea

The Baltic Sea is an area with near-perfect preservation conditions for shipwrecks. Although only 55 m average depth, the deepest point is 459 m. The most famous ship-

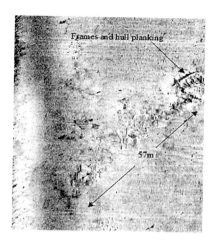

Sidescan image of the *Themopylae* wreck (ProMare)

wreck of the Baltic Sea is undoubtedly the *Vasa*, which is now on display in Stockholm. *Vrouw Maria* is another well-known shipwreck. This Dutch wooden two-masted merchant ship sank off Finland in 1771 loaded with precious artifacts such as works of art belonging to Catherine the Great of Russia (Hagberg et al., 2008). Less known are the numerous shipwrecks that a Russian nonprofit organization has discovered in the Finnish Bay in recent years in almost perfect conditions.

In Sweden a consortium was formed as early as 1983 to own and operate a small observation-class ROV system to locate and document cultural remains under water. The ROV *SeaOwl* was equipped with scanning sonar, a one-function manipulator arm, underwater positioning, and video and still cameras. It was used to document several wreck sites in Sweden and Norway (Kaijser, 1994; Westenberg, 1995). Also, during the investigation of the royal ship *Kronan*, a small Phantom observation ROV was used for diver observation (Einarson, 1990).

In early 2002, the Swedish Navy submarine rescue ship HMS *Belos* was on a routine exercise in the middle of the Baltic Sea when the sidescan sonar caught a strange-looking wreck. The on-board ROV was sent down 100 m to the bottom. Despite bad visibility near the bottom, the crew got a spectacular sight on their television monitors: an old ship standing upright with its two masts standing and bowsprit perfectly intact. The reason for the sinking is a mystery, since both hull and rigging are intact.

The ship is 26 m long and the masts rise about 20 m, with perfectly preserved platforms about halfway. The rigging is clearly a brig, with yard sails on both masts and a gaff sail behind the rear mast. It is specifically a snow

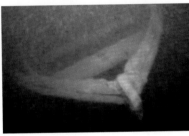

Baltic Sea shipwreck (ASK Subsea)

brig, since there is a separate minor mast for the gaff sail, just behind the main mast. The snow brig was a common rigging for minor ships during the entire eighteenth century. The figurehead on the wreck is a fantastic gilded horse with human hands clasped under its belly and a fish tail. The gunwales (shipsides) have deck-level gunports, and at least one gunport seems to have been decorated. But there are no guns to be seen. Several planks and possible decorations of the sterncastle have fallen off, perhaps because they were attached with (now rusted) nails, whereas other planks were attached with wooden treenails.

The ship seems to be a minor naval or postal ship, or maybe a private or royal yacht. Skulls from at least two crewmen lie on the deck, which is unusual, since casualties usually float away during sinking. On the video, one of the skulls on the deck seems to have an axe buried in it. Could the ship have been taken by pirates or mutineers and then sunk to remove all trace?

Several additional well-preserved deepwater shipwrecks have been discovered in the Baltic Sea in the past few years. For instance, more than forty shipwrecks have also been located during the planned Nord Stream gas pipeline route survey from Russia to Germany. Changing environmental conditions in the Baltic Sea may lead to the destruction of this unique heritage, so more detailed survey efforts should be initiated.

Norwegian Waters

The first Norwegian attempts to use underwater technology in marine archaeology were made in the 1980s. The first experiments were carried out by Nævestad (1991), who wanted to test the suitability of the then emerging technology in the management of cultural remains under water. He systematically used and evaluated several different systems, including sub-bottom profilers, sonar systems, underwater positioning systems, and the Swedish ROV *SeaOwl*.

With its strong technological background, NTNU has been mainly responsible for the development and use of new methods and technology in Norway (Jasinski and Søreide, 2001, 2008). In 1992 the Institute of Archaeology at NTNU began preparing a new academic educational program in archaeology, with maritime archaeology as one of the most important elements. This program was started in autumn 1994. At the same time, the Institute of Marine Technology established another educational program in underwater technology. The immediate goal of this program was mainly related to the exploitation of petroleum resources on the Norwegian continental shelf.

The fact that the two institutions at the same university were working with maritime aspects eventually led to close contacts among the archaeologists and technologists. With the desire for underwater research applications outside the petroleum field and need for new technologies within marine archaeology, the collaboration between the Institute of Archaeology and the Institute of Marine Technology became a reality in 1993.

Maritime archaeology is particularly important in Norway since the Norwegian coastline of more than 21,000 km is enormous by European standards. In addition, the maritime way of life has always had a crucial impact on Norwegian cultural development (Christensen, 1989). Through the joint efforts of archaeologists and technologists, the goal was to develop and use underwater technology to gain access to archaeological sites that could not be investigated by diving archaeologists—sites deeper than 30 m according to Norwegian work safety regulations.

Investigations near Agdenes

The first deepwater project carried out by NTNU was an investigation of King Øystein's harbor near Agdenes

The remarkably well-preserved harbor constructions at Agdenes (Pål Nymoen)

(Jasinski, 1995). The harbor is on the south side of the mouth of the Trondheim fjord, approximately 40 km northwest of Trondheim is one of only a few medieval sites in Norway where wooden constructions are preserved. Several features are present in three zones: on land, on the beach, and on the seabed. The harbor was probably built for political and military reasons and used as a strategic military defense post and one of a series of warning sites in the king's system of defense.

The harbor is mentioned several times in the oldest Norwegian historical records, the sagas. According to the sagas of the sons of Magnus, the harbor was built during the reign of King Øystein (1103–23). In comparing his achievements with those of his brother, King Sigurd, Øystein claims that he built the harbor while his brother was on a journey from 1108 to 1111. Later King Håkon Håkonson (1217–63) built bulwarks and restored the harbor (Sognnes, 1985).

The investigation of King Øystein's harbor started in 1773, but the deep seafloor was investigated for the first time by NTNU in 1993. The steep and rocky terrain made it impossible to use acoustic sonar equipment, and therefore the area had to be systematically inspected with ROV-mounted cameras. A differential GPS was used to position the research vessel, and the positions of the ROV and objects found in the survey were established with an underwater positioning system. An anchor and several logs from the harbor construction were discovered in deep water.

Trondheim Harbor

NTNU is situated in Trondheim, the third largest city in Norway. Trondheim celebrated its thousand-year anniversary in 1997. Ever since it was founded by King Olav Tryggvason in 997, Trondheim has been one of the most important Norwegian cities. For many years it was both the trading and political center of Norway. After King Olav Haraldsson was killed in a battle to Christianize the country in 1030, and later declared a saint, a cathedral was built in Trondheim—thus also making the city a religious center with the archbishop and a large pilgrimage.

Activities related to shipping must always have been important to this city on the fjord, and it would have been visited by thousands of ships over the centuries. It is thus very likely that the harbor area contains many traces of this maritime past. Harbors in general are considered areas of high archaeological potential. During an investigation in the Stockholm skerries, for example, a sidescan sonar from EG&G detected between fifty and sixty shipwreck sites, half of them previously unknown (Ekberg, 1997). But until recently only one historical shipwreck was known from the Trondheim harbor area—that found in the late 1960s and investigated in 1975 (Fastner et al., 1976).

In 1995 the Institute of Archaeology received a tip from a scuba diver regarding a shipwreck site. He had found bulkhead frames and wood on the seabed, which he thought originated from an old shipwreck. The diver

Trondheim harbor (Fjellanger Widerøe AS)

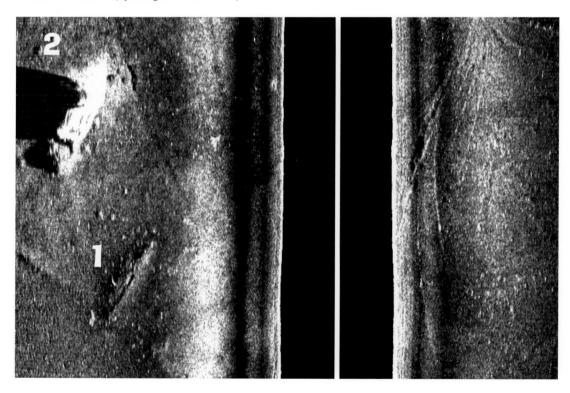
Sonar image with historical wreck site (1) and a modern wreck (2) in Trondheim harbor (Fredrik Søreide, NTNU)

ROV at the bow of the Trondheim harbor wreck (Fredrik Søreide, NTNU)

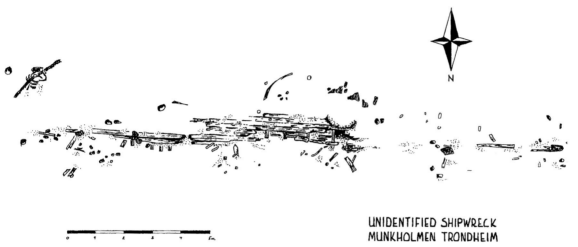

Site plan of the wreck site in Trondheim harbor (S. Carpenter)

had been in the area around Munkholmen, an island in the harbor. The depth is more than 60 m in the area, so the diver did not have time to observe much, but he was certain that it had to be the remains of a wooden ship (Jasinski and Søreide, 1997).

Since archaeologists from NTNU are permitted to dive to only 30 m, a Hyball ROV was used to locate the site. Using Kongsberg scanning sonar and a video camera, the wreck was located in less than one hour. The ROV was also equipped with a Kongsberg SSBL underwater positioning system, which positioned the ROV and objects found in the survey. Together with the DGPS position of the research vessel, this information was stored in a computer-based information management system, which was used to keep track of the areas that had already been searched and where the various artifacts had been located.

The wreck site was recorded with the ROV-mounted video cameras, and a piece of the wood was brought to the surface using the ROV's manipulator arm. The sample was dated by radiocarbon analysis, which showed that the ship is from the seventeenth or eighteenth century. Official Norwegian archives thus suggest that this site could be *Den Waagende Thrane* (Waking Trane), which is known to have sunk in Trondheim harbor in 1713. This is one of very few written statements regarding shipwrecks in the harbor area. The wreck was selected as an experimental site for testing new methods, and a major investigation was carried out over the next few years (Søreide, 2000).

In 1996 a sidescan sonar survey was carried out in this area to image both the shipwreck and the surrounding seabed. This survey was only the second attempt to locate cultural remains in the harbor. One previous sidescan survey, in the early 1980s, tried to find known modern wreck sites and the historical wreck that had been investigated in 1975 (Fastner and Sognnes, 1983); it detected a possible unknown shipwreck but could not confirm it at the time. The 1996 survey utilized Klein System 2000 sidescan sonar. It became clear after examining the results

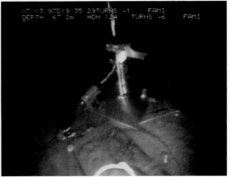

ROT system on land (left) and underwater (above) on the shipwreck in Trondheim harbor (Fredrik Søreide, NTNU)

that up to three additional possible wrecks could be situated in the same area, only a few hundred meters from the first wreck. These positions were investigated by ROV in the spring of 1997, and two sites were in fact unknown historical wrecks; the other turned out to be rubbish. The two wrecks were video-documented and measured.

A sub-bottom profiling survey was also carried out in the area to locate the three wreck sites. The experiment was done to assess the capabilities of sub-bottom profiling systems to detect buried archaeological material. The TOPAS system used in the survey is a high-performance scanning seabed and sub-bottom profiling system developed by Kongsberg. This experiment showed that it is possible to locate known archaeological material with sub-bottom profilers.

In the summer of 1996, a new investigation was initiated to better document the site believed to be *Den Waagende Thrane*. Flying the ROV to both ends of the site, the team used the Kongsberg SSBL underwater positioning system to establish the length of the site at around 20 m, with an accuracy of approximately 50 cm. Many artifacts were also measured and positioned. To confirm these acoustic measurements, a measuring rod was also used. The rod was moved around with the manipulator arm, and many distances were established by examining the rod in relation to the hull structure in the video images. A site plan was later made on the basis of these measurements and the video from the 1995 and 1996 field seasons. The plan shows the visible remains on the seafloor. Elements of a hatch can be seen near the center, and parts of the stem are still standing up in the front. By the stern there are traces of ceramics, yellow bricks, and lead. Several pieces of wood were taken to the surface by the ROV to confirm the radiocarbon age estimate.

After the 1996 investigations, it was decided that more objects were needed to establish the identity of this wreck, which could not be confirmed by examining only the structure and the limited number of objects visible on the surface. Thus, in the spring of 1997 this wreck was partly excavated by ROV—a first in Norway. Parts of the stern area were excavated by airlift and an ROV-mounted advanced manipulator arm. The airlift system consisted of approximately 100 m of flexible pipe from the ROV to the research vessel on the surface. An air inlet approximately 50 m beneath the surface was fed with low-pressure air from the surface, and the force created by the airlift could then be used to remove the sediments (silt) covering the wreck. The ROV pilot could control the process by pointing the excavation nozzle in the desired location with the manipulator arm.

The excavated area revealed a layer of boards probably originating from the deck. Several of these boards were removed by the manipulator arm, and the excavation continued beneath the boards until another layer of boards and deck girders was discovered. During excavation, several different objects were found; glass, pieces of ceramic, lead, iron, wood. Some of the large objects were picked up by the arm, placed in a collection basket, and then brought to the surface. Small objects were usually drawn up to the surface along with the sediment in the airlift system, so the silt that came to the surface was examined to retrieve small wreck fragments. The ROV *Argus* used in this project is a small work-class ROV, demonstrating that it is not necessary to use a large ROV or large research vessel to carry out underwater excavations in fairly deep water.

Again the shipwreck and objects found were measured and positioned using both the Kongsberg SSBL underwater positioning system and photogrammetry. On the basis of wreck images from the video camera or a still camera, three photogrammetry systems were used to measure distances between objects. Two reference objects with known dimensions, a rod and a cube, had been placed on the site to enable calibration of the systems. The results of the photogrammetry surveys were well within 10 cm accuracy, indicating that this method can give good results quickly. It was therefore possible to refine the site plan further based on these results.

The advantage of ROVs is their flexibility and ability to move around, which suits them to search for and document cultural remains under water. But during an excavation, an ROV's ability to move around on the site is not really important, especially on continuous wreck sites. In fact, moving and using an ROV in a restricted excavation area is more likely to be a disadvantage, since its thrusters disturb the silt and may damage remains. Remotely operated tools (ROTs) may be both more suitable and more cost-effective excavation tools.

A prototype underwater archaeology ROT was constructed and tested on the site. It consisted of a metal frame with an advanced manipulator arm, a video camera, and lights. The site was first relocated and documented with a Hyball ROV, before the research vessel was positioned over the wreck site. The frame was then lowered and successfully positioned on the wreck. It was then possible to document the area around the robot using the video camera and to take samples with the manipulator arm.

This project showed for the first time that it is possible not only to search for and document but even to excavate deepwater cultural remains using remote sensing and remotely operated equipment. It also introduced the concept of the ROT, which has since been instrumental in the development of a complete deepwater archaeological methodology. It became clear that this simple robot could have been much more advanced, for instance by adding rotational/lateral movements to the arm and camera or introducing excavation equipment and collector tanks.

Haltenpipe Marine Archaeology Project

In the late 1980s the Norwegian oil company Statoil began exploring the oil and gas reservoirs on Haltenbanken in the Norwegian Sea. As a result of finding the Heidrun oil and gas fields, Statoil and Conoco planned to build a new gas pipeline from the offshore continental shelf to the shore. After several seasons of sonar and topographic seabed surveys, a potential pipeline route, expected to be the route of fewest physical obstacles and lowest cost, was selected. This route crosses over a rugged seabed with depths of over 300 m and has a total length of about 250 km.

During the initial surveys through the coastal approaches, at Ramsøy fjord, between the islands of Smøla and Hitra, it became clear that the route ran close to a small bay where a Russian ship was wrecked in the eighteenth century. Cannon and other objects had been found in the bay earlier, but the hull had never been discovered. The cannon were situated in shallow water (3–26 m). At 32 m the seabed falls quickly to a depth of approximately

250 m and then becomes flat and sandy. Here, parts of the hull could be situated on the seabed, where they might be destroyed by the pipeline.

The Institute of Archaeology at NTNU claimed that according to Norwegian law investigations were required to establish whether the pipeline might damage cultural heritage. Statoil had not previously experienced such considerations on its pipeline projects farther south in the North Sea and at first attempted to deny responsibility for marine archaeology, but it later decided to cooperate and include the extra costs in the project (Hovland et al., 1998). Thus a marine archaeological survey was financed by Statoil in 1994 and carried out by NTNU (see chapter 8 for more detail).

The project had two major goals: to investigate the shallow part of the site containing parts of the ship's cargo and cannon; and to find and document the remaining parts of the hull and cargo, believed to be situated at depths of 250–280 m, and determine whether the pipeline could damage parts of the wreck.

Given the considerable depth, the size of the area to be investigated, and the difficult topographic and weather conditions in this part of Norway, the project raised several technical and methodological questions (Søreide and Jasinski, 2000). Technology that had not been used on underwater archaeological sites in Norway before would be needed (Jasinski et al., 1995; Søreide and Jasinski, 1998).

Historical Background   The Russian pink *Jedinorog* (Unicorn) was built in 1758/59 at the Solombalskaya shipyard in Arkhangelsk. The ship was a naval transport, approximately 39 m long, 9.6 m wide, with keel deck height of 3.8 m. The ship was carrying twenty-two cannon for defense. Ships of the pink type were built by order of Tsar Peter the Great from 1715. In Arkhangelsk, the first pink was built in 1741 and the last, of a series of four, in 1782. *Jedinorog* belonged to the third series. In total, twenty-eight of these vessels were built at this shipyard (Jasinski, 1994).

On only its second journey, the ship was on its way from the Kronstad navy base near St. Petersburg to the shipyard in Arkhangelsk loaded with an unknown number of cannon, lead, anchors, and other commodities, to be used to fit out new ships. At this time, Kronstad and Arkhangelsk were the two most important harbors in northwest Russia. All traffic between these two harbors had to go along the Norwegian coastline.

A pink, similar to *Jedinorog* (NTNU Vitenskapsmuseet)

On October 16, 1760, the ship had reached Smøla island off central Norway when a storm arose. Two days later the three masts had all broken and the ship was drifting in the dangerous waters of the Ramsøy fjord out of control. On the following night, the ship ran ashore at Sæbu island and broke up. It was crushed against the rocks and most of the cargo fell into the sea. No one seems to have survived the wreck, but twelve sailors had been put ashore at Smøla on the sixteenth.

The story of the ship was kept alive by the inhabitants of Sæbu, Hitra, and Smøla, but over the years some elements were added and some removed until the identity of the ship could not be said for certain, and it became difficult to differentiate between the truth and added elements. After more than two hundred years, the story was on the point of being forgotten.

In 1970 a local historian named Arne Stene published in a local newspaper the story that the ship wrecked on Sæbu island (the grounding site is called the Russian Neck) was the Russian ship *Voronov*, and that this wreck had taken place in 1730 (Thanem, 1970). His story was

Jedinorog site and vicinity
(NTNU Vitenskapsmuseet)

based on years of research into local traditions. The ship was supposedly on its way from Arkhangelsk to England with seven hundred Russian emigrants when it was surprised by a storm and wrecked. Most of the emigrants drowned, but some survived, only to be killed by local people. One woman was said to have survived and later to have become the ancestress of a well-known local family.

The alleged ship name and the year of the event in this story were probably inspired by a cannon found on the beach after the wreck. On this cannon was the inscription "Voronez 1730." However, more than one hundred cannon were recovered from the shallow part of the site in 1877 by a Trondheim salvage company. These cannon had the inscriptions "Voronez 1716" and "Olonets 1715," denoting the place and date of their production rather than the name of the ship. The cannon recovered in 1877 were later sold to two British companies, Rath & Raundrap and Bailey & Leetham, according to sources in the state archive in Trondheim.

After the publication of this article, Stene received information from the Naval Museum in St. Petersburg indicating that there was no record of a Russian shipwreck in 1730 in their archives. But a Russian transport vessel, named *Jedinorog* and piloted by Vasilij Bulgakov, had been wrecked in that area in 1760. Stene decided that this had to be a different shipwreck and published a new article in 1971, again arguing that the ship must be the Russian vessel *Voronov*.

In 1974, Eilert Bjørkvik analyzed the traditional stories and compared these with the written sources found in official Norwegian archives (Bjørkvik, 1974). According to the official archives, a Russian ship had run ashore on

Sæbu island on the night of October 18, 1760. This had been reported by a local named Henrik Dons to the bailiff because the Russian sailors who had been set ashore on Smøla some days before the shipwreck had come to his house and taken (unlawfully, according to him) all the goods he had salvaged from the wreck. Dons also mentioned that when he traveled to the island on the nineteenth he found several dead bodies, which were later buried, and a woman who was alive when she was found but died a few hours later. The surviving Russian sailors also gave a statement saying that they were put ashore on Smøla on the sixteenth to repair a barge and otherwise confirmed the events described above. From this Bjørkvik concluded that *Jedinorog* was certainly the ship wrecked on Sæbu in 1760.

It is interesting to note the differences between the written sources and the local tradition: different year of the wreck, different identity of the ship and mission, and differences concerning a possibly surviving woman as well as Russian sailors later killed by the local people.

With the introduction of scuba diving, a few objects were recovered from the grounding site in the 1960–70s by amateur divers and archaeologists, but the site was more or less left untouched until the NTNU project was initiated in 1994. As a part of the assessment, Oleg V. Ovsyannikov of the Russian Academy of Science searched the central archive of the Russian Navy in St. Petersburg and found not only the written statements concerning the shipwreck but also the original line drawings.

The Russian statements are generally the same as the Norwegian sources. There are, however, some small, interesting differences in the declarations given at the Russian inquiry into the loss of the ship. When the twelve sailors who had been put ashore on Smøla island to repair the barge noticed that their ship had lost its three masts and later disappeared, they had waited for better weather and then rowed out to find the ship. On the other side of the fjord they had discovered wreckage. Two dead Russians were lying on the beach, bound together. Other members of the crew were found later; some were missing a foot or an arm and some even their head. The next day the sailors returned to the site only to find people collecting goods from the wreck. When protesting against this, the sailors were threatened and nearly killed. In the end, the twelve sailors had to stay on Smøla for another eight months until they could be transported to Denmark and then back home on another Russian vessel.

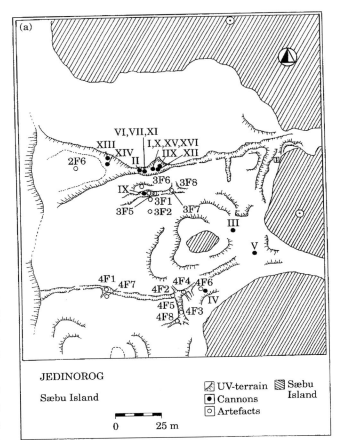

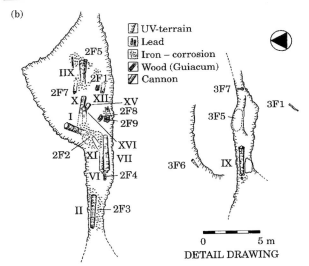

The shallow grounding site on Sæbu island (S. Carpenter)

Shallow-water Investigations The marine archaeological investigations in shallow water were carried out for several reasons. Even though the historical sources seemed to lead to the conclusion that this was the wreck site of the Russian pink *Jedinorog*, which sank in 1760, it was important to confirm this through a site investigation. It

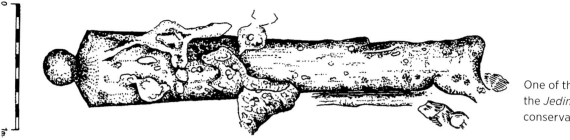

One of the cannon from the *Jedinorog* site before conservation (S. Carpenter)

A Russian five-kopeck coin from 1757 found at the grounding site. This side shows St. George; the obverse shows the initials of Tsarina Elisavieta Petrovna (J. Fastner).

was believed that recovered artifacts would confirm the identity of the ship. The goal was to document the site systematically and thoroughly. Even after several years of diving activity on the site, not to mention the salvage operation of 1877, a remarkably large number of objects were still lying on the seabed. Unfortunately, artifacts had been taken illegally by sport divers who frequented the site, and in 1991 even a cannon was raised.

The work on the site was initiated in the summer of 1994 and continued on the shallow part in the summers of 1995 and 1996. To document the site thoroughly, diving archaeologists mapped and measured all artifacts on the seabed. In addition, artifacts were photographed, video-filmed, and drawn before being raised to the surface. This work resulted in a site plan.

A total of seventeen cannon were found and documented, with inscriptions that confirmed the vessel was Russian. The most common finds apart from the cannon were several pieces of guaiacum wood (*Guaiacum officinale*) carried as cargo and considerable amounts of lead. The wood was a useful commodity, especially for ship equipment such as tackle. The lead had several applications, mainly as supports on wooden gun carriages. In addition, artifacts such as gun flints, coins, glass, and two ship weights were also found and documented. One of the ship weights has an inscription showing the city weapon of Moscow and the year 1756. Several of the coins were minted later than 1730, proving that the ship could not be *Voronov*. In 1995 and 1996 a trial trench was excavated to reveal a fairly thick cultural layer. Among the objects recovered were several gun flints and fragments of porcelain, glass, iron, and wood.

This first systematic marine archaeological investigation of the site, supported by Norwegian and Russian archives, proved beyond doubt that the ship is *Jedinorog*.

Deepwater Investigations   The main purpose of the marine archaeological investigations was to locate and document parts of *Jedinorog* that could be destroyed by the pipeline. It was believed that the remaining parts of the ship and cargo had sunk and fallen off the steep underwater cliff starting at approximately 32 m. Because this cliff does not level out until 250–280 m, large wreck parts might be found on the flat, sandy sea bottom, not far from the pipeline track, their positions depending on the effect of the current on their descent to the sea bottom.

When Statoil was made aware of this possibility, a multibeam survey was conducted in the area to locate possible wreck parts. This survey revealed four anomalies on the sea bottom. At such depths it was obviously impossible to use divers. NTNU owned a small observation ROV, a Sprint 101, equipped with a video camera

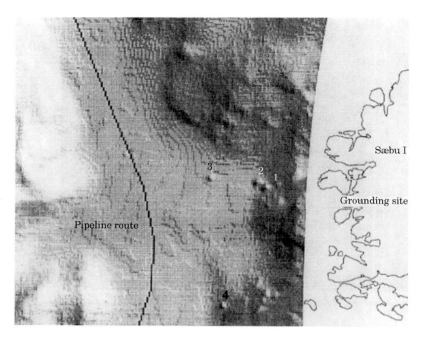

Multibeam image from the Haltenpipe project area. Note the four anomalies. (NTNU Vitenskapsmuseet)

and Kongsberg scanning sonar, which was used to locate objects from the wreck. In the summer of 1994 it examined the area around anomalies 1 and 2 but found only small wreck fragments such as wood and lead.

The seabed surrounding the four anomalies is flat and sandy, so it is relatively easy to locate objects lying on top of the seabed with sonar. When an object is located by the scanning sonar, the ROV can be flown to this position and the object studied by video camera. The objects most commonly found were, however, not wreck parts but smaller stones and piles of seaweed. Closer to the underwater cliff the terrain turns stony with small areas, galleries, and crevices filled with sand. In these areas, and in the cliffside itself, it is almost impossible to distinguish wreck parts from the surrounding rocks with sonar. In these areas only a visual search using the video cameras would help.

A new survey was carried out in 1995. This concentrated on the area around the third anomaly, but again only fragments of the wreck were found (some wood and lead). The Sprint ROV had now been replaced by a Hyball observation-class ROV equipped with a video camera, still camera, a one-function manipulator arm, and Kongsberg scanning sonar. In addition, a Kongsberg SSBL underwater positioning system was used to position the ROV relative to the research vessel. The research vessel was positioned with differential GPS.

During the 1995 season, the computer software program VETIS (Vehicle Tracking and Information System) was used to keep track of the research vessel and ROV and display their positions in real time on digital maps (Søreide et al., 1996, 1997). This made it simple for the ROV pilot to navigate and supported the team with an overview of the areas already investigated and those not yet covered. In addition, this system can be used to log information such as position, text, pictures, and video images of objects found. All of this information is stored in a database and can be replayed later and linked to the video from the investigations through time references. The VETIS database now contains information about areas searched for wreck parts, parts found, and where found (including position, text, pictures, and identification numbers), making it an excellent form of documentation supported by the video recordings.

Because nothing had been found to account for the large anomalies, a sidescan sonar survey was carried out in the autumn of 1995 to confirm the results of the multibeam survey and to better establish the position of the anomalies. This sonar survey could not confirm the existence of the four larger anomalies. Instead it indicated a large number of smaller targets scattered all over the bottom, some of them close to the anomalies from the multibeam survey.

In 1996 a SeaOwl 507 observation ROV was used to investigate the area around anomaly 4. This ROV examined the area from shallow water (20 m) above anomaly 4 and down to the anomaly at 260 m. The area around anomaly 3 was also reinvestigated, since there were some

SOLO work-class ROV (Fredrik Søreide, NTNU)

The operations room on board *Seaway Commander* (Fredrik Søreide, NTNU)

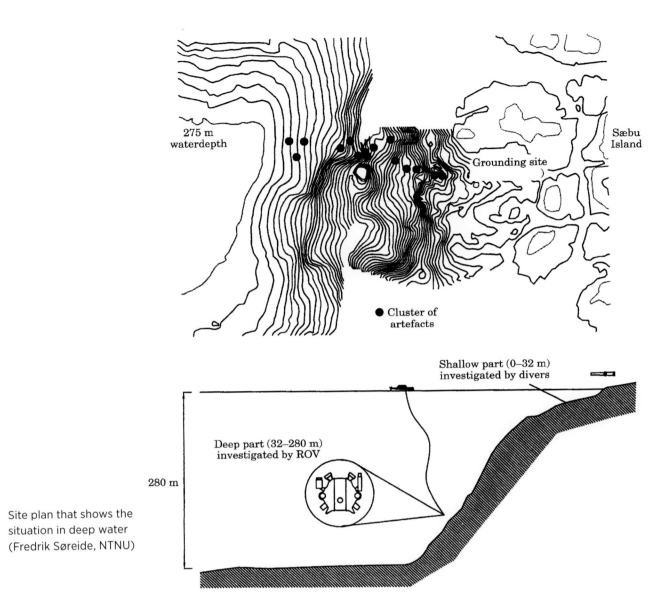

Site plan that shows the situation in deep water (Fredrik Søreide, NTNU)

new anomalies in areas not examined before. However, this survey also failed to discover large wreck parts. The Hyball ROV was used to survey an area from the shallow grounding site down to about 80 m. This area revealed several artifacts (wood, lead, iron) at 50–80 m.

Given the large amount of sediment in the area and the possibility of landslide from the underwater cliff, it was thought likely that wreck parts could be buried beneath the sediment. An investigation with sub-bottom profiling equipment was carried out using TOPAS, a high-performance seabed and sub-seabed inspection system with a unique scanning parametric sonar system and electronically controlled beam that collects detailed three-dimensional or vertical two-dimensional topographic and seismic profiling data. The system can be used to generate a real-time display of the seabed and underlying structure and thus detect wreck parts partly covered by sediment. At this location the system identified twenty-nine possible wreck positions on or just beneath the seabed. In the areas with stone, or stone covered by only a small layer of sediment, the system could obviously not penetrate the sediment and was unable to locate anything.

In addition to the sub-bottom survey, a magnetometer survey was carried out. However, since the magnetometer had to be towed high in the water column due to the complex terrain, no anomalies could be detected. Finally, seven sediment samples were taken with a grabber at predefined positions on the seabed and analyzed for metal content including iron and lead. The seven areas analyzed were considered the most promising for finding

Long wood log (ca 80 cm) at a depth of 88 m. The wood was identified on recovery as *Guaiacum officinale* and was one of several such logs found. The logs were carried as cargo. (NTNU Vitenskapsmuseet)

wreck parts. Compared with a neutral sample and most of the other samples, two of the samples had a lead content that was twice as large. These samples came from near the main anomalies.

On the basis of results from the previous years and those from the sidescan sonar, TOPAS, magnetometer, and bottom sample surveys, several promising positions were selected and investigated in June 1997. A SOLO work-class ROV installed on board the offshore diving vessel *Seaway Commander*, operated by Stolt Offshore, was used in these investigations. The ROV first investigated the targets already identified as the most promising. It identified them as either stone or piles of seaweed/garbage, and larger wreck parts were not found.

A grid survey was then carried out. The ROV searched systematically along predefined lines along the seabed, using both scanning and sidescan sonar. A few wreck parts were discovered during this survey, including some objects from previous seasons. The sidescan sonar data were analyzed, and possible targets were thereafter investigated by flying the ROV to their locations. All objects on the seabed (including stones, piles of seaweed, garbage, and parts from the wreck) with a height of more than 20–30 cm were located and examined. The objects found were registered with a description and positioned using Stolt Offshore's in-house software package. Only a few smaller wreck fragments were found. Based on this survey and the work from previous seasons, it was concluded that there were no large wreck parts on the deep part of the site, close to the pipeline.

It was then decided to investigate the steep bottom from the deep area and up to the shallow grounding site. Several gridlines were drawn up, but some parts of the cliff are very steep, making it impossible for wreck parts to lodge, and in other areas there are galleries and crevices where it would be fairly likely for wreck parts to lodge. Accordingly, the researchers attempted to follow the natural paths wreck parts would have followed on their descent down the cliff. Some of the gridlines contained no material from the wreck, but along one line in a depression connecting the shallow part to the deep part the ROV found a large number of artifacts, such as wood from the ship and the cargo and lead and iron objects. When this area was later investigated more thoroughly, several previously unnoticed smaller and larger artifacts were also found. Some were partly buried beneath sediment, some had been mostly hidden by small rocks, and others were lying in the open. The artifacts found in 1996 with the Hyball ROV at 50–80 m were located in the same area. By using a seven-function Schilling manipulator arm mounted on the ROV, several artifacts were brought to the surface, the largest being a 75 kg piece of guaiacum wood. All objects were positioned and documented with video and still photography before recovery.

The ROV was also equipped with a subsea excavation system. This was used to excavate an area at a depth of 257 m where several objects were deposited in the same place, some nearly buried in sediment. The system used was a Tritech Zip Pump, an integrated eductor-based system designed to pump mud, sand, or gravel. There are no moving parts on the dredging side of the system; the power is derived from a stream of high-velocity fluid that creates a low-pressure region behind the suction nozzle—thus preventing archaeological objects from being destroyed by the system. During excavation, the excavated sediments were not collected, only moved a few meters away from the excavation area. It would, however, have been possible to use collector tanks. The flow could also be reversed and used to blow away sediments. Both methods were used, but because of quick blockage the sediments were mainly blown away. No additional objects were discovered beneath the sediments.

Conclusions This project showed that difficult underwater topography presents additional problems to searches for cultural remains on the seabed. Steep and rocky topography makes it difficult to use sub-bottom profilers, magnetometers, and sonar equipment. Such terrain makes it almost impossible to distinguish between cultural remains and the surrounding rocks and constitutes problems for towed equipment. Such terrain therefore restricts all but visual search.

This three years of work achieved several goals: It proved beyond doubt that the wreck is the Russian pink *Jedinorog*. Events prior to and after the wreck have been reconstructed, and new knowledge about the ship itself has been gained. It has also been established that the ship broke up and that parts of the ship and cargo were deposited in shallow water. Much material can still be found in fairly thick bottom layers, especially in crevices and other natural deposits. All objects lying on the seabed have been documented and some recovered.

The rest of the ship and cargo fell off the underwater cliff and followed one particular route down the cliff side. Material from the wreck was deposited all the way down the cliffside to a depth of 280 m. There are, however, no large wreck parts on the deep, flat part of the seabed below the grounding site, and no parts close to the pipeline. Most important, therefore, this site investigation established that the oil company could lay the pipeline, assured that it would not damage the wreck on the seabed. Finally, the project was a significant contribution to deepwater marine archaeological methodology and paved the way for the most advanced deepwater archaeology project to date, the Ormen Lange project (Bryn et al., 2007; Jasinski and Søreide, 2004, 2008; Søreide and Jasinski, 2005, 2008).

Ormen Lange, the First Deepwater Shipwreck Excavation

The Ormen Lange field is in the Norwegian Sea, 100 km northwest of the central Norwegian coast. It is Norway's largest gas field and was proven through drilling by Norsk Hydro in 1997. The development of the field includes installation of a subsea production system, which will be piped directly to an onshore processing and export plant. The gas will be transported by the world's longest subsea export pipeline, 1,200 km from Norway to Easington, U.K. When it reaches full production, the field will meet 20 percent of the U.K. demand for gas.

The latest edition of the Norwegian Law on Protection of Cultural Heritage regarding marine objects was written in 1978. It clearly states that marine archaeological investigations must be carried out prior to construction work on the seafloor. These investigations are to be performed by the proper authorities, and the developer is responsible for the costs of such investigations. Overall, the law states that cultural remains on land and under the sea are to be protected in the same way. This principle was clearly established by the Haltenpipe investigations.

Norsk Hydro presented existing survey data from their proposed pipeline routes, but unfortunately those data were inadequate to detect, with reasonable certainty, the

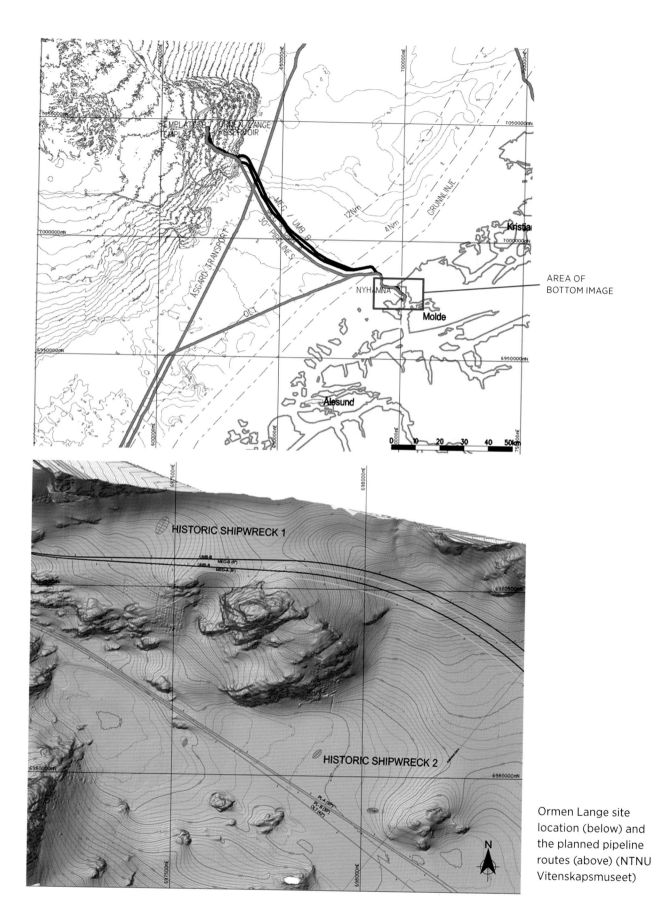

Ormen Lange site location (below) and the planned pipeline routes (above) (NTNU Vitenskapsmuseet)

presence of archaeological material in the proposed pipeline corridors. Although multibeam and sidescan surveys had detected eight modern shipwrecks in the pipeline corridors, the data were seriously flawed. The frequency used had been too low and did not provide sufficient across-track resolution to detect discrete targets in the pipeline routes. The sidescan sonar systems used also had a horizontal beam angle too great to resolve archaeological material along- and across-track adequately.

In addition, the altitude at which the sidescan sonar was towed was too great and reduced the occurrence of "acoustic shadows," which are of paramount importance when attempting to detect targets of small relief on the seafloor, particularly of an archaeological nature. But perhaps the greatest error in conducting these surveys was that the sonar system was apparently towed directly down the center of the proposed pipeline corridor. This led to the complete failure to detect objects in the actual pipeline corridor. Sidescan sonar should not be towed directly down the center of a target area, for there is a considerable blind spot beneath the sonar that is greatly augmented by excessive tow altitude, insufficient vertical beam angle, and insufficient dynamic range in the sonar system itself.

As is discussed in later chapters, recommendations for proper archaeological surveying with sidescan sonar include utilizing a modern sonar system and a digital data collection system. In addition, the sidescan sonar should be high-frequency (not less than 300 kHz), have high dynamic range, be towed at an altitude no greater than 10 m off the seafloor, and be set to a range of no more than 100 m to ensure detection of archaeological material.

Further justifying a separate marine archaeological survey, the pipelines were to pass through an area with a rich maritime history. Because of its geographic location, the area is listed on the Norwegian Directorate for Cultural Heritage's list of areas with special priority for marine archaeology and protection of underwater heritage. The area has been visited by Norwegian and foreign vessels since the Middle Ages in connection with rich herring fisheries, general trade, timber trade, and naval operations. The closest city, Molde, approximately 40 km to the south, has been a trade center since 1742. The wreck site also borders Hustadvika, a stretch of water with a reputation as a graveyard of historical and modern ships. It is mentioned several times in written sources as a place of maritime tragedy caused by storms or bad navigation.

Prior to this project there were twenty to thirty known historical shipwrecks in Hustadvika and surrounding areas.

Norsk Hydro and NTNU therefore agreed that NTNU would carry out a new marine archaeological survey in selected areas of the pipeline route (Bryn et al., 2007). During this survey in 2003, the team utilized an ROV to locate shipwrecks in the pipeline corridors. The ROV was flown along the centerline of the pipeline routes. High-resolution scanning sonar and sidescan sonar were used as the primary survey instruments. Sonar images were interpreted online and used to locate potential targets. When an interesting target was located, the ROV was flown to it and the target inspected using the ROV video cameras, while the DP survey ship was holding position.

The first survey resulted in the discovery of one modern iron-hulled shipwreck, numerous coils of wire rope, and several large steel objects that would pose a direct threat to the pipeline installation. The most important discovery, however, was a historical shipwreck situated close to the planned pipeline route.

The stern area of the ship was first identified on sidescan sonar as a dense scatter of glass bottles indicative of the mid- to late eighteenth century. A cursory examination of the bottle scatter revealed that there were probably more than a thousand bottles of varying morphology visible on the surface and clearly a significant number of bottles partially or completely buried to the south-southwest of the main site. The bottles were most likely cargo of the ship, along with stoneware containers that may have carried wine, beer, champagne, cognac, and brandy. A closer inspection demonstrated that the site was a historically significant wreck in a state of good preservation. The site contained thousands of artifacts, some exposed on the seafloor surface and others buried under several centimeters of sandy silt that moved across the site in tidal currents.

The visible wooden structure was approximately 30 m long, and the overall shipwreck site was deemed to be more than 40 m in length. The structure sits on the seafloor along a line running northeast-southwest, from bow to stern, respectively. The natural terrain of the site ranges in depth from approximately 165 m at the bow to 170 m at the stern.

The ship was splayed open through decay of the deck timbers and upperworks. Present on the bow of the ship were four lead hawse pipes through which anchor cables

Sidescan mosaic image and detail of the Ormen Lange wreck site and surrounding area (NTNU Vitenskapsmuseet)

would have been worked and which would have been at the very bow of the ship, to the starboard and port of the bowsprit and pulpit. Clearly visible in the bow were the massive cant frames and stem timber and possibly the remnants of major timbers such as the apron and keelson in a good state of preservation.

Abaft, or astern, there was an expansive area of ship timbers, some exposed and others partially or completely covered with sediment. This stern section of the wreckage contained innumerable artifacts such as porcelain, stoneware, pewter, glass, wood, brass/bronze, and iron.

Documentation   The complex terrain around the wreck site made it almost impossible to adjust the pipeline corridor away from the site. Norsk Hydro attempted to use an alternative corridor, but a new sidescan sonar survey revealed another historical shipwreck site in the middle of this route. Thus, additional investigations of the wreck site were necessary. During 2004 and 2005, advanced technology and new methods were used to document and excavate the site, making it the most technologically advanced underwater archaeology project ever undertaken.

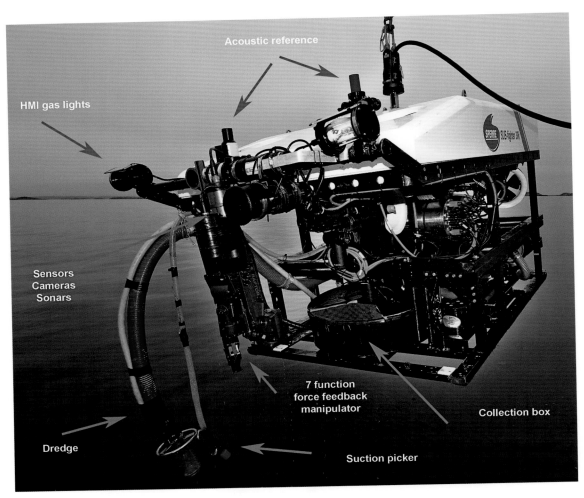

The specially developed marine archaeology ROV for the Ormen Lange project (F. Naumann)

When it became clear that extensive documentation and even excavation of the site would be necessary before the pipeline could be installed, it was decided that NTNU's extensive experience developing deepwater archaeological methods and investigating sites worldwide would be used to design a specially suited marine archaeology ROV with the necessary tooling to document and excavate the site (Søreide et al., 2006). A purpose-built Sperre Sub-Fighter 30K ROV system was therefore designed and constructed. The team also decided to use another smaller ROV system as backup, for video documentation, and for tasks that required simultaneous use of two ROV systems. Both ROV systems were operated from the research vessel *Cehili*, which was anchored in a four-point mooring system over the shipwreck site.

The area surrounding the site was initially surveyed with ROV-mounted sidescan and multibeam sonar to construct a detailed bathymetric profile of the seafloor, to determine the extent of the site, and to minimize the impact of the pipeline installation. A Kongsberg scanning sonar system was used in a downward-looking mode to create an even more resolute acoustic model of the site and seafloor. The data were used to form contour charts of the shipwreck site.

The visible parts of the shipwreck and area surrounding the wreck were then surveyed with video cameras to establish the full extent of the site and locate artifacts that may have become disassociated with the main wreck site over time. Detailed visual inspection of a 400 by 800 m area surrounding the main hull structure revealed 179 man-made artifacts in the vicinity of the shipwreck, the majority of which were modern. The inspection also revealed a spread of shipwreck-related artifacts to the south and southwest, downhill from the main site.

To complete this survey, the ROV was equipped with seven high-resolution video cameras, including HD and

Cehili: on site (top); in the operations room (middle); and from the ROV (left) (F. Naumann)

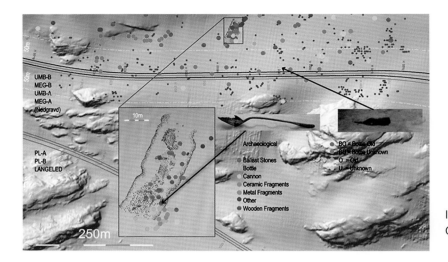

Interpretation of man-made artifacts from Ormen Lange (NTNU Vitenskapsmuseet)

3CCD broadcast cameras, as well as 1600 W HMI gas lights and 1250 W halogen lights. Paired lasers were used to introduce a scale in selected video images to obtain measurements in situ. Various measurement rods were also positioned on the site.

The ROV was also equipped with a photomosaic camera developed by WHOI, which was used to collect images for photomosaics at various stages of the investigation. Because of the rapid attenuation of light under water, the only way to get a large-scale view of the seafloor is to build up a mosaic from smaller local images. The mosaic technique is used to construct an image with a far larger field of view and level of resolution than can be obtained with a single photograph. The photomosaics were created by flying the ROV over the archaeological site at a constant altitude with the camera pointed parallel to the site in a true plan or elevation view. After the site had been completely covered, the collected images were processed in a software program that joins image borders so that the edge between them is not visible (Ludvigsen and Søreide, 2006).

In many cases archaeological material is completely buried beneath the sediments and cannot be located by visual aids. The area surrounding this wreck site was also surveyed with state-of-the-art sub-bottom profilers to determine the full extent of the site buried beneath the sediment. To collect suitable data, the ROV was equipped with closed-loop control that enabled it to operate automatically along programmed survey lines at constant altitude. A human operator would be incapable of doing this with the required accuracy. An array of LBL transponders was installed around the wreck site and used for positioning of the ROV and as the main control parameter for the closed-loop control together with a Doppler log and various other motion sensors. These data were fed into a specially developed software package to output the necessary control signals to the ROV.

Three different sub-bottom data sets were collected. The data were reviewed in native format without conversion to ensure that there was no degradation in resolution. Targets were selected only if they were single point sources or contiguous series of anomalies or diffractions, regardless of amplitude, in the near surface sediment and not deeper than the preglacial-postglacial sediment boundary. Any features or anomalies that were obviously of geological origin, with or without seafloor expression, were not selected as targets. Any anomalies that were the result of acoustic, electrical, or environmental noise in the data were also excluded from selection whenever possible. The data were plotted in ArcGIS over the available bathymetry and sidescan sonar mosaic and correlated with cultural material and modern debris to determine the origin of the anomaly in its respective data set.

Based on the sub-bottom results, the extent of the main shipwreck site beneath the sediment and not visual by visual inspection was postulated to extend south-southwest from the southernmost visible artifact on the seafloor, and the site extent was also enlarged to the west and east. The results of the sub-bottom surveys farther from the main shipwreck site proved more difficult. A few areas were selected for further study on the basis of the distribution of clusters of sub-bottom anomalies, where all sub-bottom data coincided. These areas were southwest of the wreck and along the pipeline routes, with priority given to anomalies lying along the pipeline route corridors and between the corridor and the wreck. In

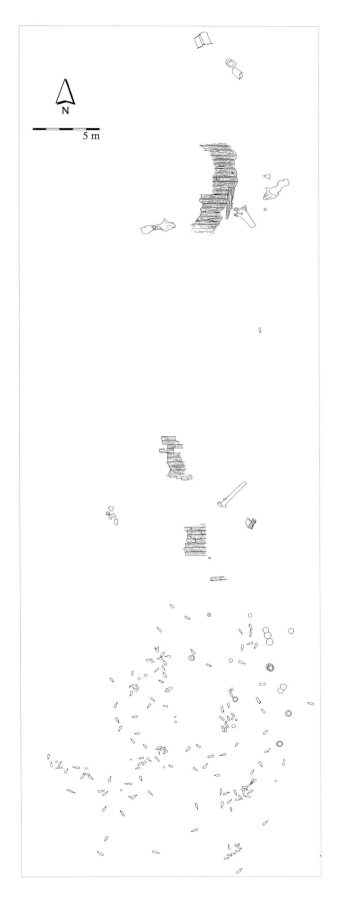

this way, any anomalies or artifacts at risk of disturbance or damage because of their proximity to the route of the pipeline corridors would be examined.

A series of test pits were then dug to correlate the presence of targets in the sub-bottom data with buried artifacts and sediment types. Close to the wreck, there was an apparent correlation between the sub-bottom data and archaeological material in the test pits. An extension of the main wreck site was clearly demonstrated during the excavation, with several meters of wooden construction uncovered to the west, east, and south of the visible structure.

Farther from the wreck the clusters were less obvious, and there was limited evidence of a direct relationship between the presence of a higher-amplitude reflector in the sub-bottom data and any buried artifact. Rather, the results of the excavations indicated a relationship between the presence of higher-amplitude reflectors in the sub-bottom data and sediment variation. High amplitude reflectors were especially common in the fauna-rich sediments and in areas with greater densities of shell fragments and small clasts. The results suggest that for small clusters of anomalies it is difficult to determine from the sub-bottom data whether the anomalies are indicative of a buried artifact or a soil attribute (high organic content, shell fragments, and clastic material). Thus, the use of sub-bottom technologies is perhaps inappropriate without other geophysical tools to investigate the shallow soils, such as resistivity measures and gradiometer surveys from suitably adapted ROVs or surface-towed equipment.

Excavation   The preexcavation documentation formed the basis for a preliminary site analysis and site plan. The results of the multibeam survey, sidescan survey, detailed visual inspection, sub-bottom survey, magnetometer survey, and test pits indicated that the spread of material from the shipwreck, visible on the surface and buried beneath the sediments, was limited. However, large sections of the shipwreck were believed to be buried beneath a thin layer of sediments, so it was decided that the site should be uncovered and partly excavated to establish the extent to which it would be damaged by the pipeline construction and to learn more about the shipwreck site itself.

Predisturbance site plan, Ormen Lange (Ayse D. Atauz)

History of Deepwater Archaeology

Ormen Lange excavation frame on land (Fredrik Søreide)

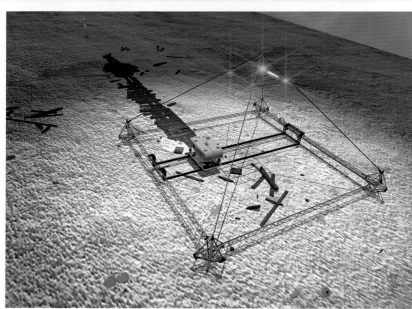

The excavation frame over the site (Brynjar Wiig)

An archaeological excavation of sunken ships is difficult, even at shallow depths. Doing it at 170 m was extremely complex. Groundbreaking technology made the world's first deepwater archaeological excavation possible.

To excavate the site with the necessary control and accuracy, an excavation support frame was developed. This 10 by 10 m steel frame was designed and constructed by the team. Suspended by a steel cable from the stern of a support vessel, it is lowered to the seafloor and placed over the shipwreck, its legs resting just outside the wreck site. The ROV then docks onto a movable platform on the frame. The docking platform can be moved in all directions within the frame by motorized cogwheels controlled

ROV on the excavation frame (NTNU Vitenskapsmuseet)

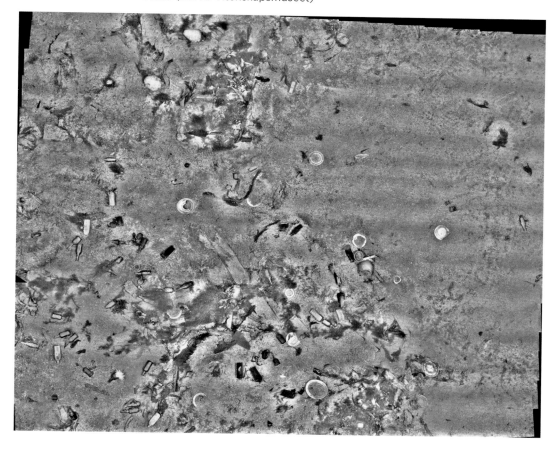

Photomosaic of the excavation frame area prior to excavation (Martin Ludvigsen)

Excavation of the stern section, Ormen Lange (NTNU Vitenskapsmuseet)

A ceramic vessel is recovered at the Ormen Lange site (NTNU Vitenskapsmuseet)

Ceramic vessel from Ormen Lange
(Ayse D. Atauz)

A specially developed marine archaeology dredge was developed to remove the sediment covering the site. It was designed to work with a water pump carried by the ROV. This pump also supplied water to jetting nozzles. The suction force could be adjusted to fit the actual sediment conditions and was successfully used to remove sediment around even small fragile artifacts. An altimeter measured trench depth. Sediments were filtered through a sediment collection basket, which ultimately recovered 250 small artifacts.

When an artifact was uncovered, it was picked up with a seven-function Kraft Raptor force-feedback manipulator arm. The force feedback function enables it to pick up fragile artifacts, but artifacts were mainly lifted by a specially developed suction picker used as the main recovery tool. It picks up artifacts with a small suction cup connected to a pump by a hose. When the pump is started, the suction cup can pick up even the most fragile artifacts. More than two hundred artifacts were recovered without damage. Some artifacts were also lifted by specially developed tooling constructed on-site. The artifacts were stored in internal collection baskets in the ROV or lifted in external collection baskets. Artifacts recovered ranged from tiny seeds and buttons to large ceramic vessels and stone plates in excess of 50 kg.

Data Management   The project contracted with ESRI to develop a software module in ArcMap to aid the recording of images and other data about artifacts as they were excavated from the shipwreck. The module consisted of a dockable window in ArcMap that could be used to input images, positions and other attribute information about artifacts as they were recovered into a database (see details on ArcMap in chapter 6).

Conclusion   The date this ship sank has not been determined with precision, since no record of its identity or loss has been discovered in spite of ongoing research in archives in Norway, England, the Netherlands, France, Spain, Portugal, and Russia. It is likely, though, that it sank early in the nineteenth century. The most modern artifact recovered is a Dutch gold coin dated 1802.

This and the date on the ship's bell are our strongest clues to the vessel's dates of construction and demise. The bell bears the inscription of its manufacture date: 1745. Even if the bell was reused, as was the custom in the

from the surface. Sitting still on the platform just above the wreck site, the ROV poses no risk to the structures and fragile artifacts scattered on the seabed and can be used to document, excavate, and recover artifacts. Positioning of the platform is based on rotation sensors of the cogwheels, backed up by high-resolution directional sonar sensors, resulting in sub-centimeter accuracy. Position input from the LBL system is also recorded. The frame allows the ROV to excavate the seafloor with great precision so that the maximum amount of data can be extracted and artifacts carefully recovered. The frame provides unique positioning control, and the combination of the specially designed ROV and the excavation frame enabled the team to conduct a systematic excavation, equivalent to a land-based excavation, at a greater depth than ever before.

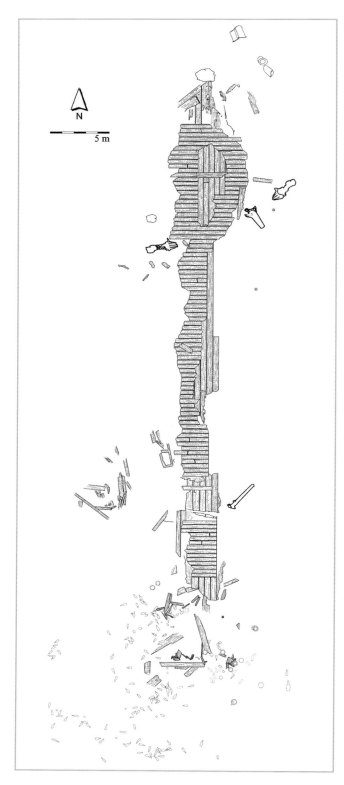

Final Ormen Lange site plan (Ayse D. Atauz)

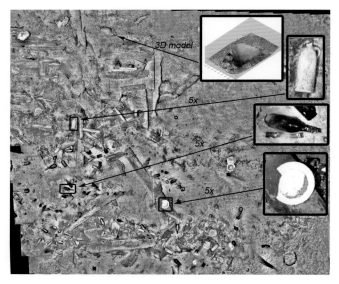

Photomosaic of the excavation frame after excavation, Ormen Lange (Martin Ludvigsen, NTNU)

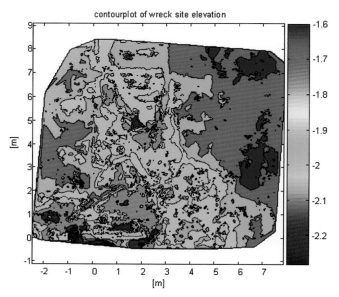

Microbathymetry of the excavation area after excavation, Ormen Lange (Martin Ludvigsen, NTNU)

eighteenth century, or reused twice in the extreme case, and even if we consider that its final owner used it for the maximum time, the date of the wrecking for the bell's last owner would be no later than the beginning of the nineteenth century. Thus, the period during which this ship was lost is likely to be between 1802 and 1810.

The artifacts discovered make it most likely that the Ormen Lange shipwreck was a merchantman that navigated along the Norwegian coast to and from Britain or

Selection of artifacts from Ormen Lange (Fredrik Naumann)

the Netherlands with a cargo meant for or originating in Russia, as suggested by the substantial number of Russian coins and other artifacts. The vessel seems to have been carrying a profitable cargo of spirits, which may have been accompanied by a load of grain, salt, or a similar perishable cargo that did not survive.

Nearly five hundred artifacts were recovered from the site, making it the most comprehensive and detailed excavation ever carried out by an archaeological institution in deep water. Research into the material collected from the site should yield the necessary clues to identify the vessel, in combination with the ongoing archival research.

The project successfully developed equipment and methods that now enable marine archaeologists to do all tasks that previously could be done only by scuba divers.

Russian five-kopeck coins
from Ormen Lange
(Fredrik Naumann)

History of Deepwater Archaeology 83

Most notably, the excavation support frame developed in combination with the specially developed marine archaeology ROV enabled the archaeologists to investigate in detail and excavate the seafloor with great precision so that the maximum amount of data could be extracted and artifacts recovered carefully. Deepwater excavation was the last frontier for marine archaeology. This new technology enables archaeologists for the first time to investigate and excavate cultural heritage in deep water with the same precision and standards as on land.

The marine archaeological project, conducted pursuant to the Norwegian Cultural Heritage Act, was financed by the Ormen Lange license and other participants in the development. The excavation was completed according to plan and in time for the planned pipeline installation in the spring of 2006. The total cost of the project was approximately US$10 million.

Treasure Hunters

Although most marine archaeologists argue that salvage of artifacts from deepwater shipwrecks by treasure-hunting companies have nothing to do with archaeology, some of the most interesting shipwreck investigations in deep water have been carried out by salvage companies (Pickford, 1993). Some of the treasure-hunting companies have had the necessary funding to initiate deepwater shipwreck surveys and recovery operations and have therefore come up with interesting and groundbreaking new methods and technology.

The following are some of the best known and most serious of the deepwater treasure hunting projects. A few additional companies are involved in deepwater search and recovery throughout the world today.

Seahawk Deep Ocean Technology

In April 1989, Seahawk Deep Ocean Technology located a shipwreck near the Dry Tortugas, about 60 km southwest of Key West. Because the material assemblage suggested a colonial Spanish origin, the ship was believed to be the remains of a member of a Spanish fleet caught in a hurricane off the Straits of Florida in 1622.

In 1622, Spain's finances were strained. The king had borrowed to help finance the war with France, and Spain's position as a world power required that the wealth found in the natural resources of the Americas be delivered to Spain. The system established for transporting the treasures of the New World back to Europe included several annual fleets of ships on standardized routes to ports in South America. Traveling in groups offered the possibility of help to a distressed ship, shared provisions in times of need, and shared mutual navigational knowledge. Routinely, the merchant or cargo-carrying vessels were accompanied by heavily armed warships that offered safe transport for the royal treasures. One of these fleets, the Tierra Firme, sailed in the early spring of 1622 for ports in South America and would later meet up with a second fleet in the port of Havana to be refitted and victualed for the return trip to Spain in the early summer.

But in 1622, problems and the weather delayed the rendezvous from the planned date of July 1 until September, which is the height of hurricane season in the Caribbean. On September 4 the Tierra Firme fleet, consisting of twenty-eight ships, sailed for Spain. Between Cuba and Florida, only one day out of port, they encountered a hurricane. The storm scattered the fleet, capsized some ships, and brought the famous ships *Atocha* and *Santa Margarita* onto the Florida keys. Nine of the ships were probably lost, two (or three) in deep water. Little is known about these ships, although evidence indicates that the wreck found by Seahawk is one of the merchant ships of this 1622 fleet (Flow, 1996b).

In the late 1960s, shrimp fishermen working near the Dry Tortugas sometimes snagged debris in their deepwater nets. The debris included a large ceramic amphora identified as a colonial era Spanish olive jar by Marx (1993). Although Marx noted the location of the site, the depth made it impossible to explore it. In 1988, Seahawk Deep Ocean Exploration, a company established by John Morris and Greg Stemm, was given the location of the site by Marx. In April 1989 Seahawk launched the research vessel *Seahawk* and, with these high-probability area coordinates, set out to locate the wreck site (Marx, 1996).

The 86-foot RV *Seahawk* was equipped with Klein 595 side-scan sonar (100/500 kHz) and a Loran C navigation system linked to a Seaquest Seatrac navigation system. The ship was also equipped with several magnetometers. The navigation system was used to generate the search grids and graphically display the position of the vessel. The side-scan sonar was towed behind the ship and created an image of the ocean floor features and characteristics up to 300 m to each side of the towfish. Skilled side-scan operators marked items of interest for closer inspection by an ROV.

RV *Seahawk* maintained its position without anchoring while the ROV was sent down to investigate interesting sites. Three ROVs outfitted with video and still cameras provided visual access and documentation. The largest and most versatile of these ROVs was a Phantom DHD2 with a depth rating of 700 m, a low-light camera, and a one-function manipulator arm that enabled artifact retrieval. A Phantom 500, with a depth rating of 200 m, and a Phantom 300, with a depth rating of 100 m, were also used. These Phantom ROVs were manufactured by Deep Ocean Engineering.

The initial visual survey was accomplished with the Phantom DHD2, which used Kongsberg scanning sonar to pinpoint the location and perimeters of the site. The ROV was positioned with an ORE multibeacon that emitted an acoustic signal received by a Ferranti Track Point II tracking system. The ROV's position could then be shown on the Seatrac system. The video survey of the site revealed an approximately 10 by 15 m wreck area, with evidence of wooden timbers, piles of ballast stones, and dozens of seventeenth-century ceramic amphoras. Although the site was in fairly good condition, it was not, as previously believed, a perfectly intact ship, 30 m high (Marx, 1993). Several 35 mm photos were taken at the site, and a bronze bell evident in the visual survey was retrieved by the ROV, allowing Seahawk to establish an admiralty claim to the site.

The wreck was located at 406 m. The seabed sloped about 6.5 degrees to the north, and the bottom current varied from 1.5 to 4 knots. Since Seahawk believed the ship could contain treasure, it decided to excavate the wreck. The company first explored the possibility of using divers. Although it is possible to dive to 400 m, the risks would be considerable and hand-excavation at that depth would not be practical. Manned submersibles were considered next but found too expensive. The best choice was a specially equipped ROV, and a salvage and excavation operation was therefore initiated in 1990 based on an ROV solution (Flow, 1996a).

Once the site was identified, the company dispatched a second vessel, the *Seahawk Retriever*, a 210-foot vessel with living facilities for a crew of thirty including scientists, ROV pilots, and technicians. The ship was equipped with a four-point mooring system to keep it in place over the site, cranes for hoisting equipment and artifacts, winches for deploying cable, equipment for communications and security, and a control module for the ROV.

A large ROV called the *Merlin* was designed and constructed specifically for archaeological excavations in deep water by AOSC of Aberdeen, Scotland. Merlin was designed to provide clear visual images in deep water, a capability to determine and record provenience of artifacts, dexterity and flexibility in methods for artifact recovery, and the maneuverability to work without endangering or disturbing the site. *Merlin*, which weighed around 3 tons out of water, was fitted with buoyancy blocks of syntactic foam to make it 300 kg positively buoyant. The system was powered by six hydraulic powered thrusters that allowed it to work above the seabed without stirring up sand or silt or crushing artifacts. In other words, *Merlin* floated and used its thrusters to push itself down.

Merlin was fitted with two manipulator arms to pick up objects as large as 120 kg, but it could also handle delicate objects without damage. These advanced five- and seven-function arms could be controlled with a miniature master arm from the control room on the *Retriever* and offered virtually unlimited range of motion to 1 m. Suction was used for both a dredge and a lifting limpet. The dredge used a venturi principle to vacuum up silt and sand. It was designed not to break delicate objects, with the criterion that it should be able to pass a wine glass through the entire system without breaking it. A complex filter system basically pulled an item out of the silt before ejecting it to the exhaust (Stemm, 1992). Material recovered via the suction dredge was funneled into a receiver, which was designed and constructed by Submarine Technology. This receiver retained all solid material and extruded seawater. The receiver was mounted on Merlin, and its contents were labeled, transferred to buckets, and sealed for transportation to a laboratory in Tampa.

To provide visual access to the site, *Merlin* was equipped with five video cameras, including a 3CCD broadcast-quality camera with surface-controlled zoom and focus functions and an SIT camera for low-light conditions to allow real-time viewing of events on the seabed. A wide-angle camera mounted high on the ROV frame was dedicated to providing a plan view overhead image of *Merlin*'s work envelope. Video images from the excavation activities were recorded with four SVHS video

ROV *Merlin* (Odyssey Marine Exploration)

Suction-based lifting limpet about to raise a ceramic jar (Odyssey Marine Exploration)

recorders. These tapes, including the data overlay and verbal comments, were kept as an important part of the excavation documentation. Two pan-and-tilt units allowed the cameras to survey larger areas of the site without moving the ROV. *Merlin* had in addition three photo cameras—one Photo Sea 35 mm camera and two 70 mm cameras—oriented so that stereoscopic photos of an artifact could be obtained in situ. Twelve 250 W quartz halogen lights and two 1500 W strobes were powered from *Merlin* (Flow, 1996a).

The ROV was operated by a crew of three from the control room of the *Retriever*. The two pilots and a data logger had been given extensive training prior to the excavation. The video was relayed to the control room via fiber optic cables, and three 30-inch video screens allowed *Merlin* to provide 180-degree views of the underwater excavation site.

To record provenience, Seahawk used a LBL underwater positioning system from Sonardyne. Four transponders, with releasable weights, were established at the perimeters of the site and provided a constant point of reference. A transponder was also mounted on *Merlin*, and a transducer on the *Retriever*, which communicated with the system. When *Merlin* sent an acoustic ping to one of the four fixed transponders, the transponder immediately sent an acoustic signal back to *Merlin*. The total time it took for *Merlin* to send a ping and receive it back was used to calculate the distance from *Merlin* to the transponder. The position of the ROV transponder in relation to the four stationary transponders could then be calculated and displayed on an electronic grid in the control room as an x,y coordinate. The depth was found with a Digiquartz depth gauge.

The underwater positioning system was also used to position receiving baskets on the seabed. Large artifacts, like amphoras, were placed in a four-plex, four plastic bins, each 110 by 110 by 75 cm, padded with heavy foam, which were arranged in a steel cradle with a lift ring. The four-plex was lowered by a mechanical winch to the site. A transponder mounted on the four-plex gave the position relative to the four stationary transponders. Smaller objects such as pottery shards and ballast stones were placed in smaller bins, which were transferred to the four-plex. Special baskets, later transferred to the four-plex, were used for retrieval of small or precious objects. When full, the four-plex was retrieved by the winch. Because *Merlin* descended and ascended daily, a vehicle-mounted container offered an alternative quick retrieval method for specific objects.

A computer system with software from Sonardyne was used to generate position coordinates of *Merlin* every five seconds. The ROV position was displayed and permanently recorded by overlay on the video screens that displayed the events on the seabed. Artifact positions relative to *Merlin* were, however, not established, and an artifact was simply given the same coordinate as *Merlin* at the time when it was first discovered.

The positioning data, along with various vehicle orientation data, were received from *Merlin*'s computers and sent to a software system developed by Seahawk. This system, the On-Line Data Storage and Logging System, was run by the third member of the crew, designated as the data logger. The system first took the received information, processed it, and sent it to the display system that overlayed the data on the video recorders. Then, as events transpired during operations in situ, the data logger entered appropriate written comments and assigned artifact numbers to items as they were first seen, photographed, and placed for storage. This information was recorded immediately to a paper log transcript and to an electronic database for further review by an archaeologist. The data-logging system was equipped with a "frame grabber" feature that enabled it to capture a video image and record it digitally on an optical disk. The paper log included a heading with dive number and date and a continuous log of time in hours, minutes, and seconds followed by the event, comments, and position coordinates.

The electronic database was stored on disks for easy access. The information in the database could be presented in a variety of ways. For example, any category of artifact could be listed. It was also possible to generate plots that showed the distribution of any category of artifacts or to place the objects over an outline of timbers that defined the perimeters of the ship. Categories could also be plotted together to show distributions over the site and relationships to the others. Total counts and frequencies of artifacts were available at any time during the operation and for postprocessing.

During the 1990 and 1991 seasons 16,480 artifacts were recovered from the wreck site. Ceramics (aboriginal ware, olive jars, glazed ware), metals (brass, bronze, copper, gold, lead, silver), organics (bone, teeth, seeds, leather, pearls, ivory, tortoise shells), glass and stone (ballast and cultural artifacts) were typical. A full inventory can be found in Flow (1996b).

Seahawk Deep Ocean Technology had located another interesting wreck site when searching for the Dry Tortugas wreck. To investigate, it formed a relationship with the Harbor Branch Oceanographic Institution, which provided access to its ships, a manned submersible, and a deepwater ROV system. The submersible, the *Johnson Sea Link*, which can be used down to 1,000 m, was sent to investigate this second wreck site at about 300 m. Harbor Branch stationed RV *Edwin Link* on the site to do some initial investigations and make a photomosaic survey with the submersible. The site was visually inspected from large viewing domes in the front of the *Johnson Sea Link*, and the submersible's manipulator arm established a grid pattern of buoys on the seabed. The photomosaic images could then be captured with a video camera and a 70 mm still camera mounted in the bow of the submersible. Many of the visible artifacts were also plotted, measured, and photographed. Some test pits were also dug with a suction pump. Overall, a variety of artifacts were discovered, but the suction pump was not able to dig very deep, and a thruster was therefore used to blow away the sand around the ballast pile. Most of the ship's lower hull is preserved. Two cannon, some cannonballs, a brass telescope, a dozen copper pots of various sizes for cooking, and about twenty other artifacts were recovered to the surface. The site can probably be dated to the early eighteenth century (Duffy, 1992).

The Dry Tortugas operation cost around US$18,000 per day, for a total of more than US$10 million. Seahawk invested in the excavation with the belief that artifacts found on the site were valuable enough to more than cover the costs of the operation. The commercial model of companies like Seahawk dictates that select trade good artifacts recovered in multiple quantities may be sold to collectors in order to generate revenue after the artifacts have been thoroughly, conserved, studied, and documented. Thus, such companies generally focus on sites likely to yield valuable salvage, yet this commercial model does not preclude sites of archaeological and historical value.

As a result, Seahawk early on decided to try to use archaeologically sound methods so that the historical significance of the site and its artifacts would be properly documented. In keeping with this policy, the company hired archaeologists to supervise the excavation, a move that was praised by some archaeologists. Other archaeologists are, however, skeptical about the commercialization of shipwreck investigations. Still, it is fair to say that this was perhaps the first serious attempt to excavate a deepwater shipwreck entirely with advanced robotics, thus providing precedent and valuable technical experience for future deepwater archaeology projects.

Odyssey Marine Exploration

Odyssey Marine Exploration was founded in 1994 by John Morris and Greg Stemm based on the belief that good business and sound archaeological practice can co-exist and thrive together. In its efforts to proceed according to this principle, Odyssey utilizes state-of-the-art technology to locate and recover deep-ocean shipwrecks around the world.

Since it was established Odyssey has surveyed and mapped more than 11,000 square miles of seabed and spent more than 9,000 hours exploring potential and

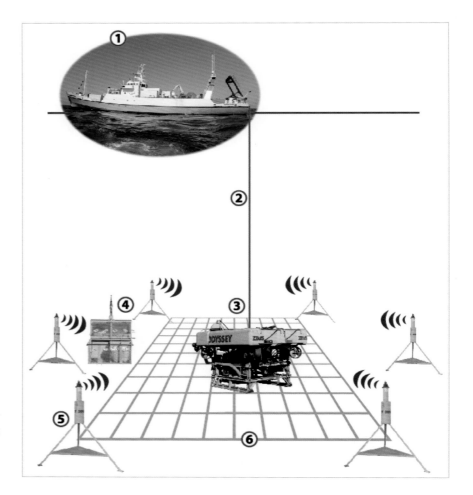

1 Surface vessel
2 Umbilical
3 ROV
4 Recovery basket
5 Positioning System
6 Gridded shipwreck site

Components of deep-ocean excavation
(Odyssey Marine Exploration)

actual shipwreck sites with advanced robotic technology. To date, the company has discovered hundreds of shipwrecks ranging from third-century BC Punic sites to U-boats and modern-day fishing vessels.

The company's most famous discoveries include the Civil War-era shipwreck of the SS *Republic* from which over 50,000 coins and more than 14,000 artifacts were recovered and documented; the *Black Swan*, a site characterized by nearly 600,000 coins, mainly in clumps scattered across an area of seabed the size of six football fields; and Admiral Sir John Balchin's HMS *Victory*, which sank in 1744. Odyssey also has an exclusive contract with the government of the United Kingdom for the archaeological excavation of HMS *Sussex*, an 80-gun English warship that sank near Gibraltar in 1694. The company has other shipwreck projects in various stages of development around the world.

Odyssey uses side-scan sonar, and in some cases magnetometers, to identify targets. During surveys, every anomaly on the ocean floor is recorded and then analyzed. The most promising anomalies (based on size, shape, location, and other factors) are considered targets and then visually inspected by ROV.

Odyssey owns and operates several vessels, and charters additional ships as necessary. The *Odyssey Explorer* is a 251-foot Class II DP ship and state-of-the-art deep-ocean recovery platform. The ROV *Zeus* is the centerpiece of Odyssey's advanced robotic archeology system. The 400 hp ROV weighs eight tons, is rated to operate at depths to 3,000 m, and is driven by eight powerful hydraulic thrusters. It mounts two Schilling seven-function Conan spatial correspondent manipulators that provide the exceptional dexterity and fine control required for delicate archaeological procedures. Recovery operations combine tooling on the ROV, cameras, and specialized computer hardware and software to record the location of artifacts in situ and throughout the recovery and conservation process. Work generally begins with an archaeological predisturbance survey, which includes a detailed photomosaic of the site. Once complete, the archaeological excavation and recovery of artifacts can begin. Details about some of Odyssey's projects follow.

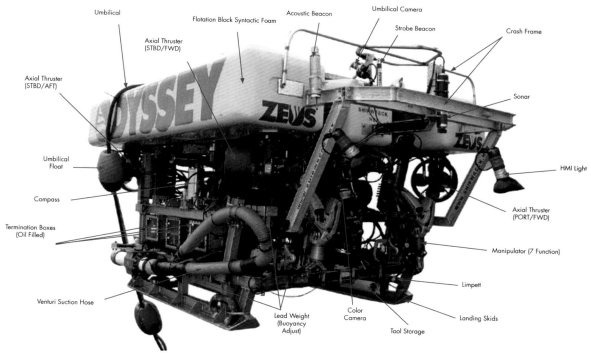

ROV *Zeus* (Odyssey Marine Exploration)

HMS Sussex This vessel was a large eighty-gun English warship, one of thirteen Royal Navy fleet ships lost in a severe storm in 1694 off Gibraltar. Documentary evidence indicates that the ship was carrying a substantial cargo of coins when it foundered. The sinking of the Sussex was observed by several eyewitnesses who later testified at a hearing held by the Royal Navy. Two vessels also witnessed her sinking and reported the loss in their logs. This data combined with other contemporary reports was essential to Odyssey in its early research and subsequent search for the wreck of HMS Sussex.

Between 1998 and 2001, Odyssey used side-scan sonar and bathymetric systems to map the sea floor and to locate potential targets east of the Straits of Gibraltar in the Western Mediterranean. The most promising anomalies were inspected visually with a ROV. During the course of these search expeditions, 418 targets were located. Some of those targets turned out to be ancient shipwreck sites, including Phoenician and Roman sites over 2,000 years old. Many were also modern shipwrecks, aircraft, geological features, or simply debris.

Only one site, located with side-scan sonar and lying at a depth of nearly 1,000 meters, presented features consistent with the HMS Sussex, including cannon distribution and size, anchors, approximate date, and location. The observed wreck above the seabed measures 33 by 12 m and is elevated 6–7 m above the surrounding sea bed.

Ten days of the 2001 expedition were spent in an attempt to identify the shipwreck remains at this site using an ROV with special tooling for uncovering and recovering artifacts. Measurements were taken and several artifacts retrieved for identification purposes.

In September 2002, Odyssey entered into a partnering agreement with the owner of HMS Sussex, the government of the United Kingdom. In accordance with this agreement, Odyssey developed an HMS Sussex Archaeological Project Plan. The fieldwork outlined in this plan comprises two phases designed to strategically examine the site's character and identity: Phase One consists of a non-disturbance Survey (Stage 1A) and Trial Excavation (Stage 1B), limited to approximately 10 percent of the wreck site area, in order to determine the orientation of the site through the exposure of key features; to reveal the level of preservation under the sediments; and to provide evidence which may help to confirm the identity of the wreck. Phase Two entails systematic and strategic excavation of the coin cargo and its immediate area. In 2006, Odyssey announced it had successfully completed the archaeological and environmental survey operations to fulfill the requirements of Phase 1A to the satisfaction of

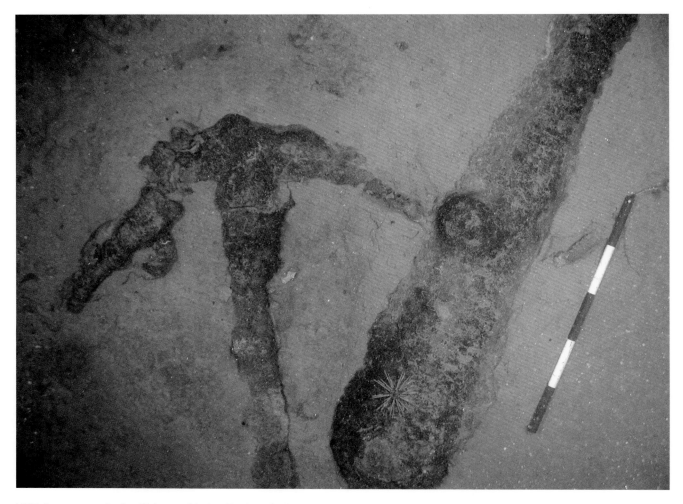

HMS *Sussex* wreck site (Odyssey Marine Exploration)

the British government and a substantial portion of Phase 1B (trial excavation of the site believed to be HMS *Sussex*). Odyssey is authorized by the British government to complete phase 1B of the project. However, due to interference by various Spanish entities Odyssey has postponed further work on the project to allow diplomatic issues to be resolved.

In July 2008, Odyssey delivered a report to the British government with the results of the environmental and biological sampling, which were submitted for extensive post-fieldwork analysis. Coring samples and other evidence indicate that the wreck site may be at least twice as large as visible on the sea floor, and could potentially include areas displaced some distance from the visible wreck mound. Only further trenching and site exploration can confirm or discount the presence of considerably more wreck material than is either evident or exposed in the limited trial trenching conducted to date.

Although the work completed to date does not conclusively identify the site as that of the *Sussex*, the evidence gathered suggests a vessel of the approximate timeframe and provenance of HMS *Sussex*.

"HMS *Sussex* Shipwreck Project (Site E-82): Preliminary Report" is available on the company's Web site, shown in this book's Reference section.

SS *Republic* The SS *Republic* was a Civil War-era sidewheel steamer. The ship was en route from New York to New Orleans, when it was lost in 1865 after battling a hurricane for two days. The passengers and crew escaped, but a reported $400,000 in specie (1865 face value) sank 518 meters to the bottom of the Atlantic.

After surveying more than 1,500 square miles, Odyssey located the wreck site with the aid of computer models that combined information from newspapers, survivor's reports, ships' logs, and other information about currents

Paddlewheel of the SS *Republic* (Odyssey Marine Exploration)

and the storm. This process laid out potential locations for the ship's sinking that reflected different combinations of wind and current variables.

During the summer of 2003 the research vessel RV *Odyssey* procured an Edgetech CHIRP side-scan sonar image that showed a shipwreck with features and dimensions closely matching those of the SS *Republic* as documented historically. An ROV inspection revealed a sidewheel partially buried by sediment, the ship's rudder, and a large field of artifacts including bottles of preserved fruit or other food with corks still intact.

Odyssey was awarded title and ownership to the SS Republic shipwreck and cargo and mobilized the *Odyssey Explorer* and the *Zeus* ROV. The work was conducted in two phases: a pre-disturbance survey and the excavation. As part of the pre-disturbance survey, individual images of small areas of the site were assembled to create larger photomosaics. These proved an invaluable for understanding the site, creating a site plan, and planning excavation strategies.

During excavation, a soft silicone rubber limpet suction device, operated by one of the manipulator arms on the ROV, aided retrieval of delicate objects. Over 51,000 coins, along with more than 14,000 other artifacts, were placed in numbered plastic containers. When filled, the containers were placed in a large, rubber-lined metal lifting basket on the seabed, subdivided to store several plastic containers. Each container and division location was numbered, video recorded, and entered into Odyssey's DataLog system.

Once lifted to the surface, the artifacts' dimensions and level of preservation were manually recorded on finds sheets and then logged onto an electronic spreadsheet. Prior to triage conservation, the artifacts were photographed on board the recovery vessel and a separate photo log and file were created. With the sub-sea and surface recording systems, Odyssey was able track the history of all the artifacts from their first observation on the seafloor to Odyssey's land-based conservation facility. Artifacts from the SS *Republic* have been on display

Gold coins from the *Republic* on the seafloor (above) and on limpet (right) (Odyssey Marine Exploration)

through fixed and travelling exhibits, and available for study since 2005.

Several archaeological papers about the SS *Republic* and the artifacts recovered are available on the company's Web site or are available in the book *Oceans Odyssey* (Oxbow Books, Oxford, 2010).

The "Blue China" Wreck   The "Blue China" wreck was discovered by Odyssey in 2003, after a fisherman had caught a glazed earthenware jar in his trawl in 2002. At the time of the discovery, more artifacts were recovered including a bowl, and a pitcher containing a glass tumbler, for the dual purposes of arresting the site in court and in an effort to identify and date the vessel.

In 2005, Odyssey revisited the site, finding substantial and ongoing modern damage, the apparent impact of modern trawl nets dragged across the ocean bottom. Odyssey chose to conduct a "rescue archaeology" excavation including documenting the site and relative positions of artifacts, creating a photomosaic, and recovering as many intact artifacts as possible before the site underwent total destruction. Most artifacts were retrieved by a soft silicone rubber limpet suction device while the ROV *Zeus* was hovering above the site.

While the ship's identity has not been established, evidence at the site suggests it is the remains of a small American coastal trading vessel. A large quantity of articulated hull structure was observed in various stages of deterioration and further sections underlay the sand substrata. A large concentration of ceramics and glass bottles was clustered near the southern end of the site, identified as the bow through the presence of two concreted iron anchors. Odyssey has named it the "Blue China" wreck after its principal cargo, consisting of blue and white pottery, later identified as largely mid-19th century British earthenware. At the date of this writing, the company has announced plans for publication of documentation regarding the site and its cargo.

Black Swan In 2007, Odyssey announced the discovery of the *Black Swan* site in an area of the Atlantic where several Colonial-era ships sank. One hypothesis associates the coins discovered as cargo from the *Nuestra Señora de las Mercedes*, a Spanish vessel transporting merchant goods and other cargoes at the time of its sinking in 1804.

Odyssey's pre-disturbance survey included more than 14,000 still images. Preliminary excavation and recovery yielded nearly 600,000 silver coins, hundreds of gold coins, worked gold, and other artifacts. This recovery possibly constitutes the largest collection of coins excavated from a deep-ocean site. Spain has filed claims for the treasure, but Odyssey continues to defend its right to a salvage award for the recovered cargo.

Mediterranean shipwrecks Many shipwrecks were discovered almost 1,000 m below the surface of the western Mediterranean during the search for HMS *Sussex*. Several of the sites are ancient shipwrecks covered by amphoras, two suggesting a Punic or Phoenician merchant vessel of the fifth to third century BC and several suggesting Roman vessels. Other shipwrecks discovered in the area include colonial period armed merchantmen, coastal traders, and modern vessels.

Commercial Model for Deep-ocean Shipwreck Investigations

The technology and larger ships needed for deep-ocean archaeology require a level of funding that is out of the reach of most academic and non-profit organizations. Odyssey has therefore developed a commercial, for-profit model, which allows for either rewards paid for work accomplished, or the sale to collectors of select, duplicative trade goods.

The company's professional mission therefore differs significantly from treasure salvage operations of the past whose sole aim was the recovery of commercially valuable items from sunken wrecks, often without regard to archaeological standards and procedures. This for-profit paradigm, while considered controversial among some in the archaeological field, has enabled the company to explore and document hundreds of shipwrecks throughout the world.

SS *Central America*

The sinking of the side-wheel steamer *Central America* was the worst American maritime disaster of the nineteenth century, claiming 425 people and more than three tons of gold. *Central America* was a wood-hulled luxury steamship with two large iron side-wheels, operating between New York and Panama. Passengers traveling with the *Central America* were mainly Californian gold miners returning east from the gold fields. In four years the ship made forty-three round-trips, and more than a third of all the precious metal found in the gold fields probably reached the East Coast on this ship. On September 9, 1857, one week out of Panama, as the ship rounded the Florida Keys it ran straight into a hurricane. It developed a leak, and when the steam boilers went out it lost its ability to hold its bow into the waves and was battered until it was filled with water. It sank three days later. One hundred passengers and crew were rescued by the *Marine*, a brig from Boston, and another fifty or so were rescued by the Norwegian bark *Ellen* the next morning. The loss of the valuable cargo touched off a wave of bank failures, for the gold was needed in New York for payments and loan support. The sinking therefore contributed to the financial panic of 1857, one of the most severe economic depressions in the United States (Kinder, 1995).

In the 1970s, three friends from Columbus, Ohio, began exploring the idea of searching for and salvaging the *Central America*. One of them, Tommy Thompson, was working in deep-sea mining, and this work exposed him to new deep-sea technologies like the SeaMARC IA, a deepwater sidescan sonar system that could be used to investigate a lot of ocean in reasonable time (Dane, 1990).

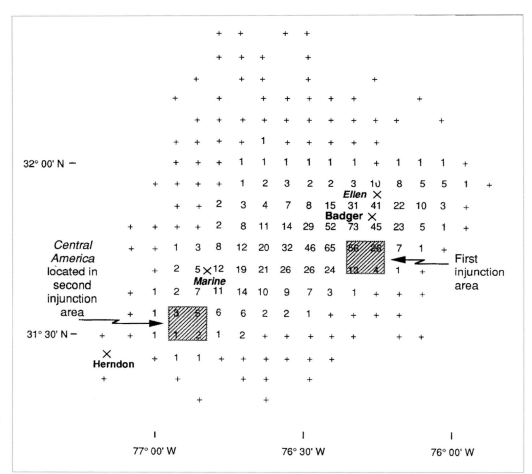

Location probability map for SS *Central America* (from Stone, 1992)

The friends began collecting all the information about the ship and events of its last voyage they could find. In 1985 they used this material to create a probability distribution of the location of the *Central America*, adapting a combination of historical, statistical, analytical, and subjective methods (Stone, 1992). Several scenarios were constructed from the available information, including position fixes from the *Central America* and other ships at the site, the accounts of the survivors, and oceanographic and meteorological data such as currents and wind. These methods were first developed during the search for the U.S. nuclear submarine *Scorpion*.

The probability distribution indicated that the ship would be found within a 3,600 km² area. Thompson and the others created a partnership, the SS *Central America* Discovery Group, and raised more than US$1 million, mainly from local business people. This was enough to use the SeaMARC for forty days, and in 1986 the search for the ship began. The SeaMARC was towed 100 m above the seabed by a long towing cable to cover the search area systematically with a minimum number of course changes. The depths were between 2,000–3,000 m, but the flat seabed and limited current in the area allowed efficient use of the wide-scan sonar. The SeaMARC has three ranges. At long range it operates at 27 kHz and covers an area of 2,500 m to each side of the sonar. Such a low frequency normally gives poor resolution, but with a stable towfish and advanced image-processing techniques the sonar system was still able to make out an item about 2 m across, enough to find the *Central America* boilers, engines, and other ironworks. The SeaMARC's capabilities spring partly from a two-body towing system that decouples the sonar sled or towfish from the movements of the ship. The SeaMARC was used to image the most interesting targets with higher resolution. The signals were recorded on optical disks and displayed on thermal paper while the sonar processing package analyzed and enhanced the sonar images.

During the search, hundreds of anomalies were discovered. Eventually one target stood out. It had to be a wreck resting upright on the seabed. A higher-resolution sonar image was made of this and some other targets, and it was apparent that the ship had similarities to the *Central America*. By mounting a video camera and a still camera to a hydrodynamic sled and trailing this more than 2,000 m deep, the first images of the wreck were taken.

Thompson decided to build an ROV to explore the site further. It would need to be capable of picking up individual items from the size of a coin to large objects, storing them for ascent to the surface, and filming and photographing the wreck. More money was raised from the partners to realize this ROV. In 1987 the first version of the ROV, called *Nemo*, was used to recover a lump of coal from the site, which allowed them to secure the salvage rights of the wreck. The group was, however, still not certain they had found the right ship, so in 1988 they used an improved version of the ROV on another interesting location at about 2,500 m. This was located in a low-probability area, but video images showed that it was a sidewheeler. A ship's bell was recovered to prove conclusively that this was the *Central America*. The persistent and carefully planned operation had worked.

Calculations of the sinking dynamics indicate that the *Central America* sank to the bottom in about eighteen minutes, landing at a speed of 2.5 m/s. This impact and various other site formation processes, including wood borers, deteriorated the site. Since the location of the gold was not known, *Nemo* had to examine the complete wreck. In 1989, when studying and documenting the site, *Nemo* discovered the first gold coins in the large debris field of the wreck. Since then, more than a ton of gold in coins and bars has been recovered from the site. *Central America* was known to have carried at least three tons of gold, which will more than cover the investment of US$12 million and the recovery operation costs of $15,000 per day. The partnership has, however, been the target of several lawsuits from individuals and companies who claim that the gold belongs to them. Thirty-nine insurance companies who paid claims after the sinking lost the first lawsuit but appealed the decision (Noonan, 1992a).

The site is fifteen hours from shore, and the 180-foot RV *Arctic Discoverer* stayed out for months at a time with a crew of twenty-five. The *Arctic Discoverer* was originally a Canadian research vessel. It was purchased by the SS *Central America* Discovery Group and outfitted with a DP system and ROV control room.

ROV *Nemo* (Milt Butterworth, SS *Central America* Discovery Group)

The *Nemo* ROV has several interesting and unusual features. Although crude in appearance, the six-ton ROV is modular in design, allowing different configurations of imaging and manipulating tools. It is usually equipped with eight video cameras, three still cameras, a stereo camera, and Kongsberg scanning sonar. The systems communicate via fiber optic cables. *Nemo* also has powerful illuminators, including four movable lights that enable high-quality images to be made. *Nemo* is operated by five operators and eleven computers. An LBL positioning system is used for *Nemo*'s navigation and positioning. The sonar transponder network is positioned on the seafloor, near the perimeters of the site, and forms an acoustic grid that allows scientists to determine the position of the ROV. This information is stored in a computer database together with other information and forms an important part of the documentation together with video and still photography. Additional positioning is by a set of reference balls carefully positioned on the wreck site (Herdendorf and Conrad, 1991).

Among the typical tooling are several manipulator arms, including a hydraulic dustpan-like device that sweeps up objects with a water jet, an arm that transmits tactile feedback to operators on the surface so that fragile objects can be lifted, and a suction picker capable of lifting single coins. A silicone injection system can solidify piles of gold coins and envelop groups of objects in fast-hardening silicone so that they can be retrieved in a single block in exactly the same arrangement as they were found on the seabed.

Nemo sits on the seabed while operating. Cameras and manipulator arms are mounted on movable booms, which enable the ROV to work in a wider area without

Gold coins on SS *Central America* site (top), and being recovered by *Nemo* (bottom) (Milt Butterworth, SS *Central America* Discovery Group)

moving and disturbing the silt. *Nemo* has hydraulically operated drawers for transporting tools to the seabed and to recover artifacts. The underwater positioning system is used to position and measure most artifacts, but a scale is often also placed in the camera's view to determine artifact sizes. A laser-based system has also been used to determine sizes. A device called SeaVac is used to vacuum the seabed to recover gold nuggets. It retains dense material, such as gold, and bypasses lighter sediments. Sediments are also excavated with a water jet, a small shovel, or an ejector. The ROV is also equipped with a propeller boom. The large overhead propeller rotates and brings clean water down to the seabed to maintain good visibility. A special device is used to cut the oak beams to get to the gold (Lore, 1990).

Valuable material and artifacts recovered from a wreck site are worth more if the operation is carried out under high archaeological standards and artifacts cataloged and certified. The gold in a gold coin is worth one amount, but as a rare coin it is worth more, and as a rare coin from a documented recovery even more. The *Central America* Discovery Group therefore decided to treat the wreck as an archaeological site, but it would be wrong to call this a proper archaeological investigation. Even though artifacts have been recovered in addition to gold, including textiles and personal belongings (Noonan, 1992b), the recovery operation was not carried out as a systematic archaeological investigation.

USS *Maine*

The USS *Maine* was a 106 m long second-class battleship built in 1888–95. In 1898 it was sent to Havana to protect U.S. interests during the local revolt against the Spanish government. On February 15 its forward gunpowder magazines exploded, killing 266 sailors, and the *Maine* sank in Havana Harbor. The cause of the explosion was not known, but U.S. media began a rather exaggerated campaign blaming Spain, and public opinion grew to declare war against Spain. And so it went. Spain lost the war, and Cuba became an independent nation.

In 1911–12 the wrecked warship was refloated, investigated, and then towed to deep water to sink again, resting forgotten for many years. The 1911 investigation found plates that had been bent inward during the explosion and concluded that the ship had been sunk by a mine. But in 1976, Admiral Hyman G. Rickover claimed that the explosion was more likely caused by spontaneous combustion of coal in a bunker.

In October 2000 a team from Exploramar discovered the well-preserved, but broken up, hull of USS *Maine*, three miles off the Cuban coast 1,150 m deep. In the same deepwater area several other wrecks were located, one a large wooden wreck that seemed well-preserved. For their wreck search and investigations, the Exploramar team used the Cuban survey ship *Ulises*, sidescan sonar, and an ROV.

Exploramar has a contract with the Cuban government, with a 50/50 split of any salvaged treasures from Cuban waters. According to Exploramar, no artifacts have been removed from the USS *Maine*.

The Grumpy Partnership

On the evening of August 2, 1993, the F/V *Mistake*, captained by Jerry Murphy, was trawl fishing nearly fifty miles south of Louisiana in 50 fathoms of water. The trawl

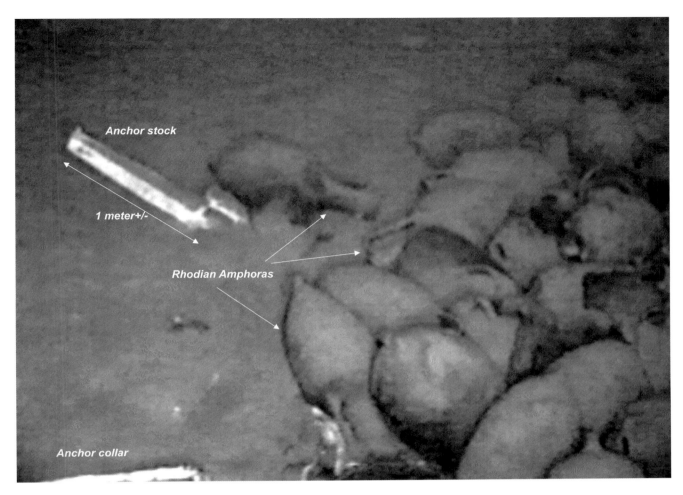

Deep, ancient shipwreck in the Mediterranean (Nauticos)

caught a "hang" on the bottom, and some time later, on being retrieved, the net was found to be damaged. When the net was cleaned, hundreds of coins were discovered, many in mint or very good condition.

At the end of the American Revolution in 1783, Carolus III, the Bourbon king of Spain, decided that the only way to save the economy of Louisiana was to redeem as much of the paper money as possible and put the territory back on a coinage system. On October 20, 1783, one of the king's most trusted captains, Gabriel de Campos y Pineda, was dispatched to Vera Cruz, Mexico, in command of the brig of war *El Cazador* (The Hunter). At Vera Cruz the ship was refitted and loaded with a precious cargo of newly minted Spanish reales and coins of other denominations. *El Cazador* departed Vera Cruz to return to New Orleans on January 11, 1784, and on its return voyage encountered a violent storm. The ship was never heard from again, and in June 1784, after an intensive search and no word of any survivors, *El Cazador* was officially listed as missing, presumed sunk with all hands.

On August 17, 1993, the Grumpy Partnership was appointed substitute custodian of *El Cazador* and financed an initial feasibility expedition to find the wreck and take pictures of items on the site surface. Grumpy also funded a subsequent salvage expedition to recover coins and artifacts. The partnership hired Marex International of Memphis, Tennessee, to stage the initial feasibility expedition and initial salvage expedition. The initial expedition successfully located the site. In 1994, Grumpy contracted Oceaneering International for further salvage services. To date the salvage expeditions have recovered some 5.5 tons of coins and numerous artifacts. Research indicates that the wrecked and abandoned vessel is indeed the Spanish brigantine warship *El Cazador*.

A WASP suit—a 1 atm diving suit with a Lloyds depth rating to 700 m—was used to salvage the site. WASP operators can walk on the bottom for six to eight hours per dive. The suit is a full metal jacket that goes down past your toes—a bullet-shaped suit you slide yourself into until your feet touch the pedals on the tube's flat bottom. Those pedals direct four propellers on the outside to let you move up, down, forward, backward, left, or right once you are in the water. Outside the water, only a crane can move you.

Nauticos Corporation

During its 1999 search in the eastern Mediterranean for the Israeli submarine *Dakar*, Nauticos Corporation discovered several interesting bottom contacts. Among the many targets analyzed, one stood out as being of potentially significant historical importance. Video and sonar imagery was collected at this site to document the target for later examination. This imagery was provided to INA at Texas A&M University to determine the origins and significance of this ancient shipwreck.

The shipwreck is Hellenistic in origin, probably dating from the end of the third century BC or a little later. The cargo was largely amphoras for wine, two to three thousand of them. Two Rhodian amphoras are clearly present near the anchor stocks at what would have been the bow of the ship. The bulk of the cargo is from the isle of Kos, close to Rhodes. Kos was famous in ancient times for the excellent wine it produced and exported throughout the region.

The shape of the wreck site is typical of ancient ships from this period. The amphoras form a more or less ovoid mound, having been stacked in the hold as many as three layers deep, and tapering longitudinally and vertically with the run of the hull of the ship. The bow is clearly identifiable, given the presence of at least four anchors.

The fact that there are several other similar wrecks in the same region is extremely interesting, for they may provide detailed information about long-distance trade over open water at specific moments in history. The fact that these shipwrecks were discovered at around 3,000 m makes them some of the deepest ancient wrecks ever discovered.

PART TWO

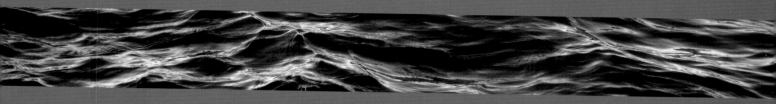

Developing a Methodology

FOUR

Location of Deepwater Sites

It is common to distinguish between a search and a survey for archaeological remains under water. A search is typically conducted for a particular site or object, such as the search for the *Titanic*; a survey is a systematic determination of the specific cultural remains within an area, for example, a survey of a pipeline route to locate all cultural remains in the vicinity.

A search should always be initiated with a period of planning in which the historical sources are studied to pinpoint the location of sites with the highest probability. This research may also include the use of computer programs to estimate the most likely search areas for ships lost at sea on the basis of prevailing wind and currents in the area. Clearly, how long a search will take depends how well the supposed site position can be narrowed down. Early planning can drastically reduce the costs, for it is less expensive than fieldwork as well as being capable of reducing search area and search time.

Collecting all available information, the planning should also try to anticipate the cultural resource material present at the site and the site conditions, for these influence the choice of search equipment and method. The most important parameters are the degree of deterioration and the metal content. The deterioration rate can be used to estimate the scope of remaining cultural material, and metal content is used as an input in magnetometer surveys. Most underwater search equipment was originally designed for larger-scale objects or structures than those sought by marine archaeologists. Archaeological applications usually require a high degree of resolution, for most archaeological anomalies are very small. Thus, the required sensitivity may go to the limit of the search equipment in question.

In shallow water, most sites are discovered by divers, and it is fairly easy to relocate these sites. In deeper water, such tips seldom occur, and historical sources are also less helpful, since most shipwrecks in deep water disappeared without a trace. Typically, then, a larger area has to be surveyed to locate the target. With increasing water depth, the complexity of search operations naturally increase too. The equipment becomes larger and more advanced, cables must be longer, and the sea conditions are more adverse. A larger ship and crew are required, and so are the costs. For these reasons, only a relatively few deepwater searches and surveys have been conducted.

These factors and many others dictate the optimal type of equipment and method for any given search, and this is the topic of this chapter (and see Kelland, 1991; Akal et al., 2004).

Seabed Mapping Technology: What Is Sound?

Sound is simply a distortion in a medium. The physical properties of the medium dictate the speed, attenuation, and other properties of the sound we hear. For convenience, sound is often described as "waves"—compressional waves or P-waves. The acoustic imaging/mapping devices of greatest interest to marine archaeologists rely on the simple reflection of sound; P-waves bounce off a target on the seafloor back to the sonar instrument and are displayed on paper or computer monitor. Coupled with global positioning systems, these sonar devices allow the mapping of large areas of seafloor with good detail, much faster than an area can be searched visually.

One of the fundamentals of acoustics is that the velocity of sound in any homogenous medium is constant

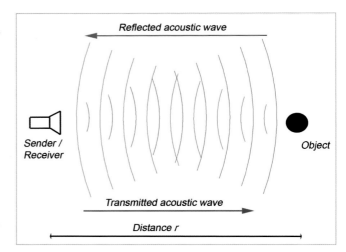

What is sound? (Fredrik Søreide)

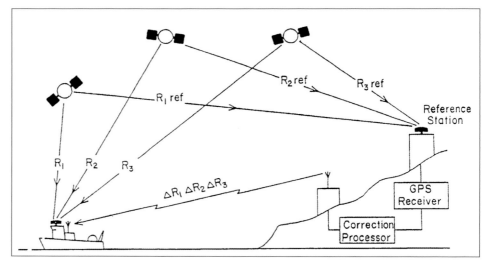

Explaining GPS (Fredrik Søreide)

regardless of the frequency of sound being propagated through the medium. The speed of sound in distilled water and at standard pressure and temperature is about 1,500 m/s. In seawater the velocity of sound varies between around 1,525 and 1,550 m/s, depending on temperature, depth/pressure, and the composition of the water. For most rough calculations 1,500 m/s is used, since it is convenient for quick math. So if we transmit a tone through water 750 m deep, the tone will take one second to reach the bottom and bounce back to the transmitter.

The most basic acoustic instrument is the depth sounder, or echosounder. In its simplest form, all that is required is a "bell" of some type (transmitter), a reasonably good means to measure time (computer), and good hearing (receiver). Modern echosounders generate a conical beam of sound and project it at the seafloor. How the sound reflects from an object, the seafloor, or different layers within the seafloor is dictated by the fundamental laws of physics. Materials that have an interface of differing impedance oscillate differently under compression and are not able to absorb or transmit the entire P-wave energy incident on their boundary; therefore the sound, or P-wave, must bounce back toward the source, or reflect. These processes can be used to create an image of the seafloor and objects lying on top of the seafloor.

Now that we have an overview of acoustics used in marine remote sensing, it makes sense to backtrack a bit and discuss the instruments of primary interest to the marine archaeologist and their practical application:

- Global positioning system
- Echosounder
- Multibeam echosounder
- Sidescan sonar
- Magnetometer
- Sub-bottom profiler

Global Positioning System

The accuracy and ultimately value of an archaeological survey is controlled to a large extent by the ability to acquire data in a systematic and predictable manner. To accomplish this task, it is necessary to navigate the survey ship precisely.

Positioning of the survey ship is achieved using radio beacons or satellite navigation systems, such as a global positioning system (GPS). Most ships today use GPS devices in which satellites moving in known orbits transmit information on their positions continuously. These signals are received by the receiver on board the ship and allow calculation of the ship's position. Depending on the region, the particular system in use, and the number of satellites used, the accuracy is usually better than 10 m.

This accuracy may not be good enough for archaeological purposes, and differential GPS (DGPS) equipment is often used. If a GPS receiver is placed on a precisely known position, it can be used to measure the difference between the known, true position and the position found by the stationary GPS receiver. This difference can be sent via radio transmitter to the mobile DGPS receivers in the area, such as on board a survey ship. The mobile DGPS receivers can then use this information on the actual errors in the GPS system to correct its signal. The resulting position accuracy is typically better than 5 m. If the reference station is within 30 km of the mobile receiver, an accuracy of less than 1 m is consistently obtained. With a thoroughly calibrated system with several land- or platform-based transmitters, it is possible to achieve decimeter accuracy.

Bathymetric Mapping

A bathymetric survey is typically carried out using an echosounder, multibeam echosounder, or bathymetric sonar. These systems are usually hull-mounted on a surface ship, but they can also be mounted on an ROV, AUV, or towed vehicle. Combined with a multitude of sensors that position the ship and the underwater vehicle, together with heave, roll, pitch, and yaw correction, a variety of software applications can be used to build a digital terrain model of a specific area with high accuracy (Ingham, 1984).

The echosounder models the seabed using single depth readings taken at intervals along survey lines whose positions are accurately known. An acoustic transducer emits a short pulse of sound at a given frequency in the water column. The pulse reflected back from the seabed is collected, and its travel time down and back and amplitude are measured. Water depth is then measured and can be presented as a depth profile beneath the ship. Therefore, if an archaeological site such as a shipwreck has a height above the seafloor, it can potentially be detected on this depth profile (Hughes Clarke, 1998).

A multibeam echosounder takes a swath of depth measurements to each side of the ship, and the depths are calculated for each point along the swath. Swath width is approximately four times the distance to the bottom. This is clearly a much faster way to map the seabed. However, each ping typically represents several square meters of seabed, and spatial resolution decreases with increasing water depth. A typical system has a horizontal resolution of 5–10 m at a depth of 300 m and 20–40 m resolution at 1,200 m. This shows the advantage of mounting the system on an ROV or AUV, which allows the system to be operated much closer to the seabed.

Even though echosounders can be used to create an image of the seabed, the resolution is usually not sufficient to detect cultural remains. Still, many modern sites have been located with echosounders, and in some extreme cases even small seabed targets can be detected.

Sidescan Sonar

Sidescan sonar is perhaps the best search tool for marine archaeological sites on the seabed. By covering hundreds of meters to each side of a moving sensor, it provides an excellent means of surveying an area of interest rapidly. Sonar uses acoustic signals to draw an image of the seabed and seabed features. An acoustic signal is sent out and received by a transducer. Sidescan sonar transmits in a sideways direction, enabling a plan of the seabed topography to be plotted as the sonar travels along the seabed. This enables the position of topographic features such as boulders and debris or archaeological sites that have not been covered by sediments to be recorded (Fish and Carr, 1990, 2000). The transducers are typically mounted on a towfish, which is towed behind a ship, or on an ROV or AUV.

Sidescan sonar emits high-intensity, high-frequency bursts of acoustic energy in fan-shaped beams to either side of the towfish. The beams are narrow in the

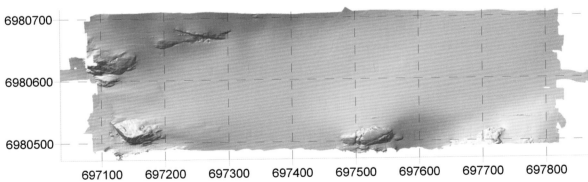
Terrain models (above and opposite) created with multibeam echosounder (Kongsberg Maritime). Examples from the Ormen Lange site (NTNU Vitenskapsmuseet).

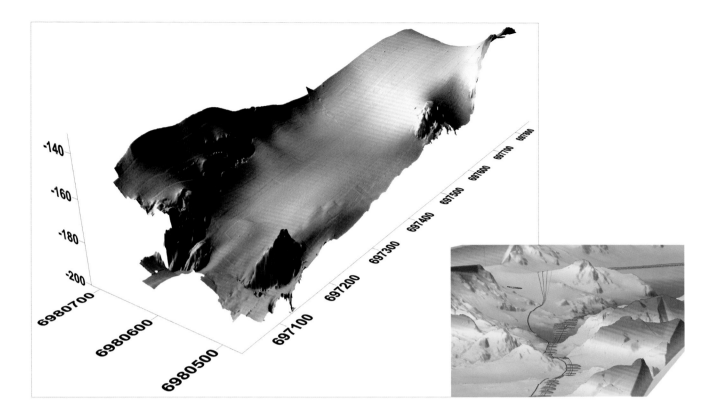

horizontal plane and wide in the vertical plane. The vertical beam angle is typically in the range of 40 degrees and is tilted 10 degrees below horizontal; the horizontal beam angle is in the range of 1 degree. The smaller the horizontal beam angle, the better the resolution. A larger tilt angle is useful when operating the towfish high above the seabed, to avoid blind spots beneath it. A smaller vertical beam is useful in shallow-water operations to minimize sea clutter.

When the sound beams project along the seabed, objects and topographic features on the seafloor produce echoes, which are received by the transducers and used to draw an image of the seabed and its features. The sonar does not measure the depth or distance but the time it takes for the acoustic transmitted sonar pulse to travel from the transducer to the target and return (Mazel, 1985). The accuracy of the sonar system depends on its ability to measure this time precisely. The range is related to the travel time by the speed of sound in water. A sidescan sonar image is then built up by laying down successive scans to form a composite image.

The acoustic signal emitted from the sonar transducers is subject to various losses, for example spreading, background noise, and absorption. These losses are higher the greater the frequency of the acoustic signal. This means that the sonar systems with the highest frequencies have the shortest range, whereas the low-frequency, long-range sidescan sonar systems are victims of more effects, for the sound pulses travel much longer distances and through different conditions. On the other hand, high-frequency sonar has greater directivity, or narrower beams, which leads to higher resolution. The operating frequency is therefore the most important parameter when you are selecting a sonar system; the choice is a trade-off between resolution and range, as can be seen in table 4.1. Moving from a 100 kHz to a 600 kHz system increases the survey time by at least five times, keeping the tow speed constant.

The resolution is, however, dependent not simply on frequency but on a combination of transmission frequency, wavelength, beam shape, and pulse length. Resolution is

Table 4.1 Combinations of sonar resolution and ranges for selected frequencies

Sonar frequency	Resolution	Range
30 kHz	Low	1,000–6,000 m
100 kHz	Medium	500–1,000 m
300 kHz	Good	150–500 m
600 kHz	High	75–150 m

Location

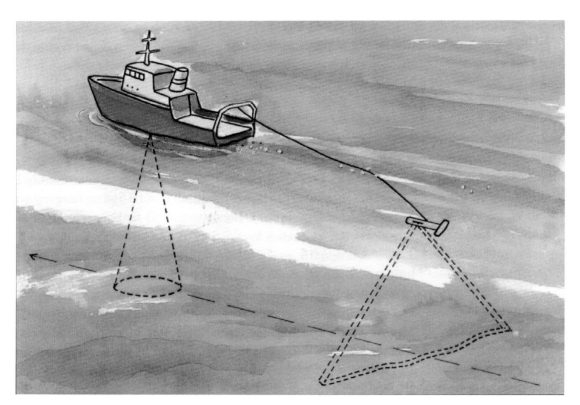

Typical sidescan sonar system, towed (above) and AUV-deployed (right) (Fredrik Søreide; Kongsberg Maritime)

ences resolution in the along-track direction. Resolution in the along-track direction means distinguishing two targets on the seabed in the direction of the tow. If two objects are separated by a distance less than the spread of the sonar beam at that range, the return signals run together and the two targets appear as one larger target. Along-track resolution also depends on the tow speed and the interval between pulses. The tow speed should usually not exceed 5 knots. If the ping rate is slow and the tow speed high, the resolution will be poor. The tow speed influences both resolution and area covered per unit of time. Thus, there is a trade-off between tow speed and resolution.

Additionally, heave and yaw distort the images, so the transducers must be mounted on a stable tow platform moving perfectly straight at a constant speed.

Instrument resolution is only part of the story. Seabed features must have a different reflective characteristic than their neighbors to stand out. When a sound pulse impinges on the seabed, the portion that is backscattered toward the source is determined, in complex ways, by the texture, slope, and material properties of the seabed. To detect an archaeological site on the seabed, the sonar system must be able to resolve the features that characterize the site (Quinn et al., 2005).

These factors have serious implications for underwater archaeological searches. The results of a sidescan survey depend on many factors, including the quality of the sidescan sonar, the underwater terrain, and the material properties of the site remains. In most cases only small objects or a pattern of small objects are visible. The material composition and geometric shape of these objects determine the reflection coefficients; for example, a waterlogged wooden artifact has acoustic impedance much like water, resulting in a very small reflection coefficient, since most of the acoustic energy passes through the wood and avoids detection. In addition, seabed sediments and the direction in which the sonar is towed over a site determine the detection rate. Even complete sites can be difficult to recognize if the tow angle relative to the site is not favorable. In some cases seabed sediments lead to spreading, absorption, and scattering of the signal, and it becomes impossible to distinguish archaeological features from the terrain, especially where the wreck parts rest on complex rock formations. In some areas remaining materials may have been covered by fouling assemblages (communities of plants and animals

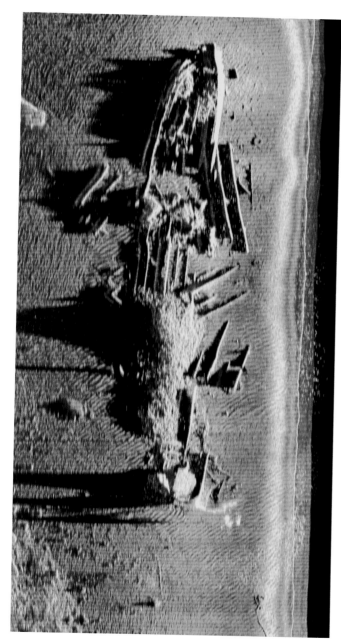

Sidescan sonar image of a well-preserved shipwreck (Marine Sonic Technology)

defined as the ability to distinguish between two distinct objects. If two objects are too close to each other, then the returning echoes from the two overlap and blend together as one target on the display. Theoretical minimum separation is one-half of the physical pulse length, so higher frequencies (short pulses) have better resolution, in some cases millimeter resolution (Mazel, 1985).

A narrow horizontal beam width is also necessary to create sharp images on the seabed. Beam width influ-

that live on man-made objects; see chapter 7), making it extremely difficult to separate artifacts from the surrounding seafloor.

When choosing sidescan sonar appropriate for a particular search operation, it is important that the system resolution be matched to the terrain and anticipated cultural remains. This is primarily a matter of frequency. In complex terrain with structure and only deteriorated remains, it can be difficult to distinguish archaeological artifacts from the surrounding terrain unless the resolution of the sidescan is very high. This is typically the case in shallow waters, near the coast. Here the terrain is often more complicated and wide sonar swaths are not possible. In such areas it is usually necessary to use sidescan sonar with a frequency higher than 100 kHz to locate the archaeological sites with sufficient resolution.

Sidescan sonar is typically towed. To be used to best effect, the towfish should be flown at a height above the seabed equal to 10 percent of the scan range being used. Thus the tow cable becomes long in deep water. Cable length to tow depth is typically 3:1 but can be reduced to 2:1 by the use of well-designed depressor vanes. In all circumstances the cable lengths are in most cases substantial, necessitating a cable winch and armored tow cables. A deep tow system is therefore heavy, expensive, and complex to use. A typical survey at 2,000 m depths requires a complete unit weighing around seven tons, including a winch with 4,000–6,000 m of cable (Wright, 1997). In really deep water it is also common to use a two-body towing configuration, with a depressor and a neutrally buoyant towfish that is decoupled from the ship's heave motions, permitting operation in bad weather (Mearns, 1995).

One should remember that the overall resolution of a sonar system also depends on its recording device. It is clear that detail is lost if we take a picture that represents 100 m of seabed and compress it down to a 30 cm print. There are, therefore, some obvious and immediate advantages to a computer display over conventional paper records. Enhancement tools can optimize image contrast and colors and improve image quality. Data can also be displayed at a variety of resolutions with the ability to zoom into a target within the sonar record. This can clearly provide more detailed information and allow more accurate measurements. Computer-based digital systems can also be directly interfaced to a ship's navigation system so that all the sonar imagery can be geo-referenced in real time and interpretative data can be logged into a database (Kiernan, 1997).

One feature of a sonar acquisition and processing system is that various filters can be applied to improve the results, including slant range correction, noise removal, and real-time removal of the water column. Slant range correction give maximum resolution, for it is used to produce a laterally consistent interpretation of the seabed by adjusting the spatial distortion caused by the towfish altitude. Water column removal and slant range corrections may, however, reduce sensitivity to objects in the water column and just beneath the towfish and are often not used in real time (Penvenne and Penvenne, 1994).

Sidescan data can also be displayed together with bathymetry and sub-bottom imagery in the processing system and used for site comparison to eliminate distortions caused by geological features.

Sub-bottom Profiling

Sub-bottom profilers are acoustic systems that can be used to penetrate the seafloor to detect buried archaeological material. These sonar systems trigger a burst of sound energy (ping) and detect the returning echo from various layers beneath the seafloor. The amplified echo signal is displayed as a function of travel time and direction. Sub-bottom profiling equipment is defined by two of its principal operating characteristics: penetration and resolution. To a significant degree, both characteristics depend on the frequency and bandwidth of transmitted pulses. The frequency of acoustic signals generated by various types of sub-bottom profiling equipment ranges from tens of hertz to tens of kilohertz.

Higher frequency is invariably associated with increased resolution and decreased penetration, so there is a trade-off between the two. High-resolution devices must be used to detect archaeological material. These systems have a range of about 1 kHz to 30 kHz. They achieve significantly improved resolution, but their penetration is limited. Typically, at 3 kHz they achieve a penetration of 30–50 m in soft materials such as clays and silts but minimal penetration in course, compact sands and gravel tills.

Sound energy transmitted to the seafloor is reflected off the boundaries between layers of different densities (hence, of different acoustic impedance). The first such boundary is between the water and the seafloor itself. As layers of clay, sand, and various other sediments succeed each other, they create interfaces from which sound is reflected. It is the energy reflected from these boundaries that the system uses to build an image.

The resolution of a system is measured by its ability to separate closely spaced objects, in other words, to detect discrete echoes returning from the interfaces between layers. The vertical resolution of an acoustic sub-bottom profiler refers to the minimum distance between adjacent layer interfaces that can be visually distinguished in the image produced by the system. A sonar system with a 10 cm resolution resolves layers that are at least 10 cm apart; layers spaced closer than 10 cm are resolved by the system as one layer. In a single-frequency system, the frequency and the length of the pulse determine the limit of resolution. In a multifrequency system, it is the bandwidth that sets the system's theoretical resolution.

A survey with sub-bottom profiling equipment is accomplished by transmitting acoustic signals through the water column into the underlying sediments and measuring the time interval between pulse transmission and the arrival of the reflected signal. Reflections occur at the interfaces between two layers of different density. Since objects buried in the sediments have a different density than the surrounding sediment, they too create a reflection. The travel time the signal uses to penetrate the sediments and return can be converted into depth to give a point-by-point profile of the position of the reflecting boundary or object. If the density of adjacent layers is too similar, no reflection occurs. The greater the difference in densities, the stronger the reflection.

Many types of cultural remains have properties with a considerably higher reflection value than normal subsurface geological structures, resulting in a strong reflection. It is therefore possible to detect, for instance, wooden artifacts, pieces of metal, and ceramics using sub-bottom profiling systems. There are also often a large number of natural objects buried in the sediments, like stones and wood, so during a survey even a small area can provide several hundred anomalies. Sub-bottom profilers do not have the necessary resolution to distinguish between a buried natural feature and buried cultural remains. Detecting a high number of sites without knowing what they are has only limited value. The suitability of most sub-bottom profiling systems for archaeology is therefore limited by insufficient resolution.

In addition, most systems operate in a single-beam mode, which means that these systems must be applied in a dense search pattern. The inability to distinguish between archaeological remains and other targets and the slow survey speed combine to make it difficult to survey for archaeological sites using sub-bottom profilers. One project in Scotland attempted to use a sub-bottom profiler to locate an ancient shipwreck. The survey resulted in several hundred possible buried wreck sites in an area of only a few square kilometers. Ongoing research may lead to new search routines based on known reflection coefficients from archaeological material and lead to improved interpretation capabilities in the future. For the time being, sub-bottom profilers are best used to survey limited areas and assess the buried content of sites already located.

Magnetometer

A magnetometer measures the strength of the earth's magnetic field and can detect variations in the field caused by the presence of ferrous material. It is sensitive to every local and abnormal variation in the earth's magnetic field. Many investigations have used magnetometers to detect and define historical shipwrecks, establishing the magnetometer as an efficient survey tool. However, a magnetometer is sensitive to both survey trackline spacing and sensor height, or distance from the anomaly source, and thus has limitations in deep water.

The intensity of a magnetic field is measured in gamma. The earth's magnetic field is usually around 50,000 gamma. Daily variations of 50 gamma and magnetic storms cause the field to fluctuate. There are also short time variations of 1–5 gamma, and magnetic inhomogeneities in the soil and scattered content of natural boulders may cause variations of up to 10 gamma. The magnetic field is also locally distorted by the presence of ferrous artifacts on the seabed. These small, spatial variations or magnetic anomalies are detected as small-scale variations in the earth's total field measurements. The detection capability depends on the sensitivity of the magnetometer. The most exceptional sensitivity is obtained by the nuclear resonance magnetometer, which can detect small amounts of ferrous metal from a long distance with a typical resolution of better than 0.1 gamma.

The proton magnetometer system creates a strong magnetic field in a hydrocarbon fluid such as kerosene by passing an electric current through a surrounding coil of wire. The spinning protons in the fluid align themselves within the temporary magnetic field. When the current is switched off, the protons realign themselves and precess back to the direction of the local magnetic field. This generates a small signal that is directly proportional to the strength of the earth's magnetic field. The precise relationship between the signal strength and field strength can be used to detect small variations in field. Cesium magnetometers operate in a similar way but have a higher sensitivity and are more tolerant of movement.

For large-scale surveys, the magnetometer is towed behind a survey ship. A magnetic coil is capable of locating an anomaly from quite some distance, but the closer the system is to the bottom the better, since a magnetic anomaly increases by eight when the distance between the sensor and the object is halved. This means that the tow cable must be long in deep water.

Towing the magnetometer as close to the bottom as possible emphasizes targets immediately under the sensor and commonly results in a series of features along the trackline, with few anomalies centered between survey lines. An alternative approach is to balance the selected sensor tow height and trackline spacing such that the magnitude of an anomaly associated with a ferrous object lying halfway between two tracklines is within 50 percent of that associated with an object lying directly below a trackline. A survey using this approach shows a more even distribution of anomalies both along and between the survey lines and is achieved with a trackline spacing equal to 2.5 times the sensor height above the seabed.

In addition, trackline spacing should not exceed 50 m, since the amount of ferrous mass on a site is typically limited and the detection capability decreases with increasing distance. For example, a one-ton ferrous mass results in a reading of 500 gamma at 5 m but only 4–5 gamma at 50 m. With a lane spacing of 30 m, all anomalies of 10 gamma or higher should be considered a possible wreck site, and the same anomaly should appear on at least two lanes (Murphy and Saltus, 1990).

The amount of metal content at a site should always be estimated on the basis of historical information. The number of cannon is usually the best indication for ships of war. Having a general idea of the target mass and thus a target gamma readout for the site being sought drastically increases the efficiency of the search, for only anomalies with the correct level need to be investigated; other likely anomalies—like pipelines, cables, or trash—can mostly be avoided. There is, of course, a risk that the signature estimate is wrong or similar to the signatures of other objects.

The magnetic anomaly from a site is also dependent on formation processes, that is, on whether the site is continuous or discontinuous. A continuous site yields a linear distribution of multiple anomalous peaks within the overall pattern produced by the remains of an intact hull, along the long axis of the anomaly pattern. A discontinuous site consists of multiple anomalies in a large area (Barto Arnold, 1996).

A major problem with using magnetometers to search for underwater archaeological sites is that it is almost impossible to interpret a magnetometer anomaly without sending a diver or ROV to investigate. The magnetic anomaly could be an anticipated wreck site but could just as well be modern waste or a pipeline with the same gamma signature. In areas with a lot of waste or high metal content in or on the seabed, it is difficult to use a magnetometer to search for cultural remains. These areas are predominantly found along the shoreline, since the deepwater areas have less waste. In cases where the anomalies are buried in the seabed, the identity of the sites cannot be determined without an excavation. This is both time consuming and expensive.

Many hundred shipwreck sites and other submerged sites have been located with magnetometers, but mainly in shallow water (Barto Arnold et al., 1999). One major survey was conducted across the submerged harbor built by King Herod's engineers at Caesarea in Israel (Boyce et al., 2004). A total of 107 line-km of high-resolution marine magnetic surveys (nominal 15 m line separations) and bathymetry were acquired over a 1 km^2 area of the submerged harbor. Underwater excavations conducted at this harbor site over the past two decades have revealed a wealth of information, but the high-resolution magnetic mapping surveys added information about what the structure looks like and how it was constructed by identifying the presence of hydraulic concrete. Magnetometers are most effective on modern sites, where the metal content is high, whereas ancient ships with almost no metal content are difficult if not impossible to detect.

Towed Cameras

Cameras can be used to search for deepwater sites, mounted on an ROV, AUV, or towed system. Many versatile, stable, and controllable systems have been designed

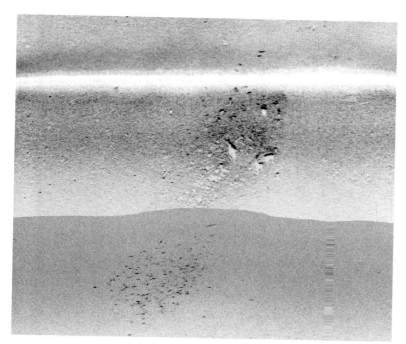

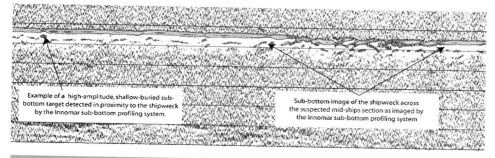

Sidescan (top), sub-bottom (middle), and magnetometer (bottom) systems: examples from the Ormen Lange site (Fredrik Søreide; NTNU Vitenskapsmuseet)

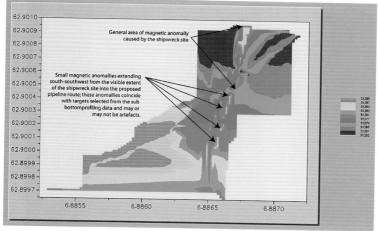

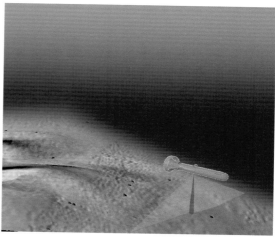

for depths down to 6,000 m and cable lengths of up to 10,000 m. They usually carry a multitude of sensors and are commonly fitted with video cameras, obstacle avoidance sonar, sidescan sonar, depth gauge, sound velocity probes, still cameras, lights, pan-and-tilt units, and transponders—in fact, everything an ROV would carry except for the thrusters. Control is accomplished by operating foils at the stern of the vehicle. The suitability of towed camera systems depends on the underwater terrain and visibility and requires that sites are not buried. Severely limited underwater viewing remains a problem when searching for sites with cameras, since the operational

Location

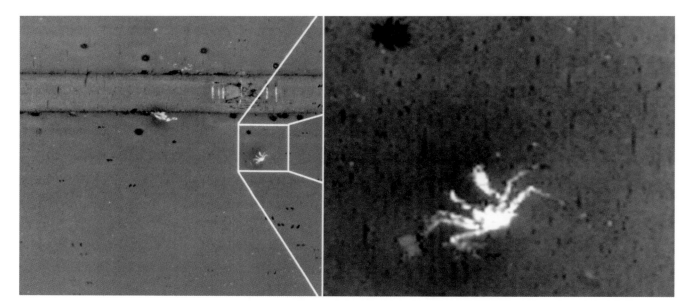

Image from a laser line scan camera

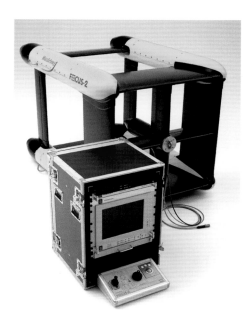

Towed camera sled (MacArtney Group)

range of most optical systems is only a few meters. The area covered by a television camera is very limited, but by combining several sensors on towed sleds they can be more effective.

Laser line scan cameras are an alternative optical system for observing the seabed (Saade and Carey, 1996). A rotating prism in a towfish sweeps a laser beam across the bottom. The laser line scan cameras build up an image from this rapidly acquired series of spots on the seabed, each point being sequentially illuminated by the pencil-sized diameter laser beam. This technique minimizes the effects of backscattering. The reflected signal is received by a detector in the towfish, and the signal can be enhanced digitally by sophisticated imaging techniques. By using up to 2,048 points per line, centimeter accuracy is achieved. Survey operations are conducted in a manner similar to sidescan sonar and produce real-time images. Mosaics and post-plots of the seabed can be generated from them (MacDonald et al., 1997).

High-power lasers allow these systems to be three to five times farther from the target object than video, so these systems can be towed at altitudes 2.5–40 m above the seabed. Survey rates obtained by laser line scan systems are therefore much higher than that obtained by an ordinary video camera and approach that of high-frequency sidescan sonar. Survey rates of 350,000 m^2/hour have been achieved. In addition, images from laser line scan cameras have much higher quality than sidescan images and are much easier to interpret because they are optical rather than acoustic. Still, because these systems are complicated and expensive, they are not often used in archaeology.

Seismic Equipment

Unlike high-resolution survey instruments, whose sensors must be towed close to the seafloor, three-dimensional seismic instrumentation consists of streamers and air-gun arrays towed just below the surface. A typical survey boat tows six to sixteen streamers that can each be up to 12 km in length. An air gun or sparker (40–250 Hz) is usually used for deep penetration to 300–1,000 m below the seafloor. Three-dimensional seismic technology has proved to be a real boon to the oil and gas industry, and with proper postprocessing the resulting data have also been useful in mapping surface faults, vents, and other "geological hazards." For example, biologists use this kind of data to predict the location of chemosynthetic tubeworms. It would be serendipitous if three-dimensional seismic data could also be used to locate shipwrecks, since this information is routinely acquired in deepwater lease blocks. Unfortunately, this is not the case. The U.S. Department of Interior's Minerals Management Services has carried out several tests and observed that the resolution is not good enough to detect even known steel-hulled shipwrecks.

FIVE

Documentation of Deepwater Sites

After the search phase has been completed, potential targets must be investigated to determine whether they are archaeological sites, to document the main features, and ideally to establish an identity. Several features may already have been established in the search phase, since the exceptional degree of resolution obtained by search equipment such as high-frequency sidescan sonar, towed camera systems, and laser line scan cameras can provide valuable information, including the type of site and approximate dimensions. But to create acceptable archaeological documentation it is necessary to send some kind of remote intervention system to the site, such as an ROV or manned submersible. This equipment must support and enable efficient and cost-effective data acquisition, including video and photo documentation, measurements, and sampling (Mindell and Croff, 2002).

Navigation and Positioning

Navigation and positioning of underwater vehicles and the positioning of site features are usually accomplished with acoustic positioning systems (Greenough et al., 1996). Long baseline systems (LBLs) provide high-accuracy positioning and measurement data. A network of transponders is deployed on the seabed around a site. A series of measurements are then made between each network transponder to calculate the distance between them. When the positions of the network transponders have been found, the relative position of an acoustic pointing device can be found by measuring the distances between the pointing device and the network transponders. The pointing device can be installed on an ROV and used to position it, enabling ROV navigation. To position and measure artifacts and site features, the ROV can place a pointing device on the target; this gives its position relative to the network. The length of a cannon can be found by placing a pointing device on each end of the cannon. In this way, artifacts can be measured and positioned relative to each other within the network and used to make a site plan. All the positions are three-dimensional and given in x,y,z coordinates relative to the network transponders. The data are transferred from the transponder network to the ship via an acoustic link.

An LBL underwater positioning system is usually said to have centimeter accuracy, since the range can typically be found with an accuracy of better than 5 cm. But positioning accuracy of the LBL systems is not determined by range measurements alone. Accuracy also depends on the number of transponders, network geometry, baseline lengths, and network calibration. Using accurate sound velocity profiling sensors, range accuracies down to a few centimeters can be obtained, and ROV positions can be calculated to within a few decimeters. The term "position" relates to both the position of each array transponder and the ROV's position, and consequently it comprises many coordinate values. The ranges from one transponder to all of the other transponders as well as the ranges from the ROV to each transponder must be measured and the accuracies of these measurements added together.

The errors can be classified as either random or systematic. Random errors vary from ping to ping, whereas systematic errors are fixed for shorter periods. If, for instance, a sound velocity calculation is wrong by 1 m/s, it remains wrong for all measurements in a limited timeframe. In addition, factors such as a roll, pitch, and heave of a ship-mounted transducer, errors from the depth

sensors in the transponders, and draught (position below waterline) and tidal movement influence the final position. The total error is the sum of the systematic error between the transponders and the random error. This adds up to the statistical sum of the errors, which is normally in the decimeter range.

To obtain the best possible positioning data, it is necessary to know the speed of sound in the water, which varies slightly from location to location. This is important since an underwater positioning system uses sound velocity to translate the travel time of an acoustic ping into a distance. Sound velocity differs slightly depending on water depth, temperature, and salinity. A value is usually taken to be reasonably constant throughout normal operating conditions, but allowances should clearly be made for its variances when undertaking detailed survey work. A sound velocity probe that measures temperature, salinity, and depth can be used to find the true velocity of sound in a given area of water.

Positioning accuracy can be improved by using smoothing algorithms such as a Kalman filter and additional motion reference data from other sensors such as a Doppler velocity log, gyro compass (roll and pitch), or a motion reference unit. A Doppler velocity log measures the Doppler shift of sonar signals reflected off the seabed to obtain the velocity of the vehicle. With these adjustments, the position data can be approved, allowing the vehicle to hover with no deviation or to maintain high-accuracy station on a survey line.

Short baseline systems (SBLs) and ultra/super short baseline systems (SSBLs) are much easier to use. They are also based on range and bearing of acoustic sound beams, but the range and bearing are measured from several hydrophones (SBL) or one transducer (SSBL) mounted below the ship's hull to a transponder mounted on the ROV. Knowing the speed of sound in water and the time the acoustic signal takes to travel from the hull-mounted transducer to the ROV-mounted transponder and back enables calculation of the distance between the two. In addition, since the acoustic signal is received by several hydrophones (SBL) or several elements in one transducer (SSBL) below the ship, the angle of the sound beam can be determined, since the acoustic pulse is received with a small phase difference. In total, this results in an x,y,z

position coordinate for the ROV transponder relative to the ship, which can be used to position the ROV, chart selected site features, and make measurements.

The advantage of the SBL and SSBL systems is the easy operation, since no transponders must be installed on the seabed, and the relatively low cost. Their accuracy, however, is substantially lower than for an LBL system. The positioning accuracy of these systems is typically 1–2 percent of slant range. Thus, at a depth of 100 m the accuracy is around 1–2 m, and if the depth is increased to 1,000 m the accuracy is 10–20 m. Thus, SBL and SSBL systems are not well suited for high-accuracy positioning and measurements on archaeological sites, but they represent a quick and cost-effective alternative to LBL.

The basic commercial systems available today come mainly from two companies, Kongsberg and Sonardyne, with several low-cost producers following suit. A Kongsberg system was utilized on the Ormen Lange project (see chapter 3), and Odyssey and other companies have used Sonardyne (Holt et al., 2004).

A few other systems have been developed mainly for marine archaeology. One of these systems is SHARPS (Sonic High Accuracy Ranging and Positioning System) (Caverly, 1988), which uses an ultrasonic device to send very short high-frequency pulses through relatively long distances in water. SHARPS is useful as a general-purpose ranging system based on high-frequency, spread-spectrum acoustic signaling. Typically a network of ultrasonic transceivers is established around the site. Divers or ROVs then carry a portable transceiver around the site. Using the pencil, the diver or ROV can outline selected objects or selectively pinpoint and record the position of discrete targets. When the pencil is positioned on a target, it emits short ultrasonic pulses (300 kHz), which are received by the transceivers at times proportional to their range from the pencil. Each transceiver then sends a "pulse received" signal to a computer, which gathers all distance data, determines the ranges, makes the necessary triangulation calculations, and displays the resulting position coordinates of the pen/object on the computer screen. Each point's x,y,z coordinate is recorded and can be simultaneously displayed.

The EXACT precision acoustic navigation system (200 kHz) is another "acoustic tape measure," providing

Opposite: documenting sites, equipment types and methods. Courtesy Brigit Luffingham and John Ramsay.

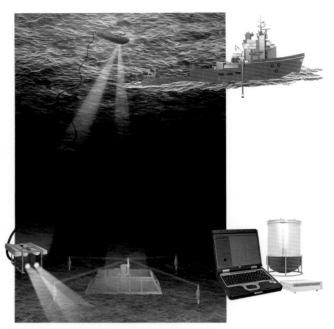

LBL versus SSBL: Ship to ROV communication (right) and ROV to transponder communication (below) (Kongsberg Maritime)

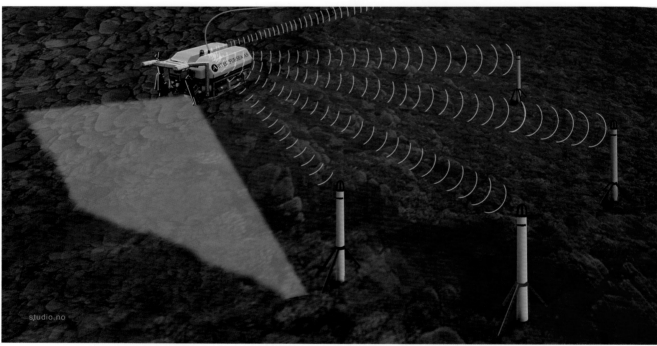

position information by trilateration. First, accurate measurements of the local speed of sound through the water are collected. Then, two or three battery-powered acoustic transponders are placed around the archaeological feature, establishing baselines. One "host" transponder is fitted to the underwater vehicle working on the site. Several times per second the host transponder interrogates the other transponders around the site by sending an acoustic signal, and the time lag between send and receive is recorded. This time lag is a range measurement: with two or more EXACT ranges, the vehicle's position can be determined. The EXACT system has been used extensively in R. D. Ballard's deepwater archaeological investigations and by WHOI and MIT.

The AQUA-METRE D100 is another local underwater positioning system that has been used for underwater archaeology (Medard, 1997a,b). It is based on the SSBL principle and consists of a measurement base, which constitutes the reference coordinates, at the top of a 2 m high mast. This mast remains underwater during the whole measurement operation.

A light portable pointer, which includes a small keyboard and LCD display panel, allows the diver or underwater vehicle to measure positions and save them in the backup memory. One base can work with up to eight pointers simultaneously. Each pointer can be fitted with a rod that allows measurement offset from sea bottom (the rod length and orientation are compensated automatically via internal inclinometers and compass).

On return to the surface, the pointer data is downloaded to a standard PC computer station equipped with infrared serial data link and the data input to a GIS system. Alternatively, the measurements can be monitored from the surface via a hydrophone. This system was, for instance, used to take measurements on the eighteenth-century wreck *Epave aux Ardoises* located west of France near the Gulf of Morbihan (Brittany). This was a relatively small ship (about 15 m long) carrying a large number of slates that led it to sink rapidly in shallow water (10 m). The measurements were concentrated on the front half of the ship, which had been made free of any slate. The main goal was to measure the three-dimensional shape of the wooden frame. Only one diver was needed to run the system during this operation, in which about two hundred points were measured along two successive dives of 45 and 30 minutes. Back on the surface, the points were transferred from the pointer to a PC computer, and the measurements were processed using CAD software to build the shapes and volumes from the measured points.

Relocating Archaeological Sites

The result of a search phase is usually a high number of targets. These targets usually have a known GPS position. The survey vessel can then be sailed to the target positions and an ROV sent to the seabed to locate the targets to document them. The survey vessel must be chosen to fulfill requirements related to deck space, crane capacity, accommodation capacity, sea-keeping and motion characteristics, transit speed, and the like. Most inshore operations can usually be completed from a small vessel of opportunity anchored over the site. Larger vessels for offshore operations are much more expensive and usually rely on automatic station keeping with the aid of a dynamic positioning (DP) system.

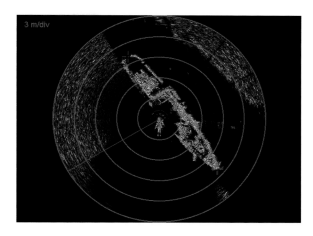

Shipwreck detected by scanning sonar
(Fredrik Søreide, NTNU)

A DP system is critical to keep a ship over a site during a deepwater investigation, so that anchoring can be avoided. The system controls the ship's position and heading. Active control of the thrusters and propellers counteracts environmental forces (wind, waves, and current) and prevents deviation from a specified point. The position-holding accuracy is typically a few meters, depending on weather condition, vessel construction, installed power, positioning reference systems, and control system. With the propellers running, however, much more care must be given to the operation, and a tether management system is usually required to avoid flying the ROV into the thrusters.

Once on the seafloor, the ROV systems must relocate the target located in the search phase. The ROV should be equipped with an underwater positioning system. Only in rare cases does the ROV find the target immediately, and a period of systematic searching of the seafloor in the target area is often necessary. Ideally, the target area is no larger than a few hundred square meters. The ROV pilot can use the underwater positioning system and other motion sensors to follow systematic search lines above the seabed while video cameras and scanning sonar are used to locate the target.

Sidescan sonar images are obtained by moving a fan-shaped sonar beam along a straight line. If we instead rotate the sonar beam to generate a so-called sector scan image, the images can provide extended vision for ROVs and can be used for obstacle avoidance, search, surveying, and positioning. The scanning sonar uses a conical beam pattern that is very narrow in the horizontal direction. As sound beams project along the seabed, the echoes from objects and topography are used to create an image of the seabed and possible targets on the seafloor. The reflection characteristics of the target determine if it can be found in the same way it would be with sidescan sonar. The scanning sonar produces an image of the seabed 360 degrees around the vehicle at a range of up to 100 m using a transducer that is mechanically scanned around the circle in about two hundred steps. A scan smaller than 360 degrees is also possible. Using high frequency, excellent image resolution is obtained. Complete images are made by stacking the sound beams. When a target is located by the scanning sonar, the ROV can fly to the location and the site can be investigated using the video cameras. Scanning sonar often works best if the seabed is flat with little structure. In areas with rugged terrain, a search must often rely on cameras only. Once the target has been located, the exact position can be found using an acoustic positioning transponder on the ROV, or acoustic pingers can be placed on the site for easy relocation later.

Search and documentation are typically two distinct and independent phases. After the survey phase, the results are processed, and targets are logged on maps and charts. Since it is difficult to determine the correspondence between features imaged with sidescan sonar and those seen in the ROV video displays, it is common to overlay ROV positions and motion in real time on maps and sonar images. By using input from the navigation systems, both the research ship and ROV positions can be displayed on top of a map or sonar mosaics of a target area with logged contacts and features. The ROV can then be flown to interesting areas by a pilot, who can see the ROV's position relative to these features. Video and other information can also be linked to features in the computer system as they appear. Several such systems exist today.

Video and Photo Documentation

The purpose of the documentation phase is to gather more information about the site. One of the most important tasks is extensive video and photo documentation, to provide a permanent record of a site and the work that is carried out. For an underwater archaeologist, vision is perhaps the most important of all senses in that it accounts for the majority of the information perceived. The remote viewing system should therefore be regarded as an extension of the human visual system. Information that is lost will seriously reduce the effectiveness of the documentation phase. The limited field of view offered by underwater cameras, especially in highly turbid water, can therefore drastically reduce the value of the documentation phase.

Video and photo documentation under water are also affected by reflection, refraction, loss of intensity, loss of color, and loss of contrast. In many circumstances, however, it is possible to compensate for these effects by using stronger lights (Sewell, 1994). On the other hand, because of backscatter—light reflected from particles suspended in water—the field of view must be reduced, and the intensity cannot always be increased by these compensating means. The effects of strong backscatter are usually minimized by positioning the light sources at an angle to the plane of the lens of the ROV-mounted cameras. But in many areas good documentation can be obtained only at a range of a few meters, even centimeters, because of turbid water. Turbid conditions are, however, usually a shallow-water problem.

During film documentation, the ROV should be docked on the seabed to avoid stirring up the silt, or be slightly positive, which means that the vertical thrusters must thrust downward to keep the vehicle near the bottom. In this configuration the propeller wash does not disturb the bottom as it would if the ROV was negatively buoyant and had to thrust upward. Camera and light booms and pan-and-tilt units are also often used to achieve optimal camera and light configurations. To obtain good images, fiber optic cables should be used. These cables are immune to electromagnetic interference and have low loss and wide bandwidth, so quality is maintained despite long cable lengths.

Underwater Photography

Digital cameras are now generally taking over as the preferred system. Most cameras for underwater photography give 5 megapixel resolution or better. Digital cameras are user friendly, the image is immediately available, and it can be stored on a computer. Images can also be sent to the surface without line loss.

The alternative film formats for underwater still photography are 35 mm and 70 mm. The larger 70 mm produces larger negatives, which is a major quality advantage.

Almost all underwater still photography is accomplished with high-energy strobe lights that provide the balanced illumination needed to take high-quality photos

ROV hovering above shipwreck (NTNU Vitenskapsmuseet)

under water. The strobe and camera must be separated by at least 0.5 m to avoid backscatter. Alternatively, the site can be illuminated by a fixed light of sufficient strength (Couet and Green, 1989).

Video Documentation

Video is the main tool of early phases of scientific documentation. Some of this material is also used for public broadcasting. Mainstream underwater scientific applications can be separated into long-range viewing, close-up detailed inspection, and general observation. Each application places constraints on camera choice. Long-range viewing requires high light sensitivity, and so intensified sensor cameras like the SIT (silicon intensifier target), ISIT, and ICCD are the best choices. For the other applications, color and monochrome CCD (charge coupled devices) or HD (high-definition) cameras are used.

SIT cameras are mainly used for navigation and observation and can operate in extremely low light conditions (1×10^{-4} Lux). They can be used without floodlights, and they improve visibility in turbid conditions because the lack of an artificial light source minimizes backscatter. These cameras are well suited for navigation, but since the picture is monochrome it is usually not the best choice for close-up imaging. CCD cameras have become the standard for all types of inspection and observation, with good-quality color video pictures and light sensitivities ranging from 10 to 0.01 Lux. To achieve broadcast-quality, 3CCD cameras are necessary. Whereas traditional analogue CCD cameras offer around 500 television lines, 3CCD typically has more than 800 lines. New digital camera formats are developing fast, and image quality continues to improve. There are standard definition digital cameras for underwater use, but the best option is high-definition cameras, which typically have resolutions of 1280 by 720 pixels or 1920 by 1080 pixels. These new digital formats also have the advantage of no transmission line losses through a fiber cable, which is common with standard video through long lengths of tether.

Video images have normally been recorded on tape, whereas most new digital systems record to hard disc. An average project usually produces several hours of recordings, since the aim is to document the complete archaeological site. Usually a low-light camera is used first to document a site from a distance and create overview images, then color cameras provide close-up documentation of artifacts and details. The quality of the results depends on the speed of the camera movement. If the camera traverses over the surface of an object quickly, the surface appears blurred when viewed on a monitor. And because video images are made up of a series of lines, most cameras have only limited capacity to resolve fine detail. This, together with the fact that video systems show two-dimensional pictures, means that the picture viewed on the monitor must be interpreted by the archaeologist. Stereoscopic video systems can be used to increase the feeling of telepresence, and the renewed general interest in three-dimensional imaging probably means that this technology will soon be used underwater.

VisualWorks is a professional suite of applications to record, archive, review, and report digital underwater video running on standard PCs with Microsoft Windows. It is designed for any type of video inspection where large volumes of video must be recorded from one or more cameras and synchronized with other data sources. Selected information can be annotated on the video recordings. Data such as date, time, depth, and position can be included and used to link the video to the other information (and time-stamped), enabling easy relocation of video sequences. VisualEvent is an event data entry tool that combines with VisualWorks to provide input of user-defined information in real time. VisualEvent is fully configurable to archaeological requirements and can be operated by an archaeologist online or offline, making full use of the fast access to digital video and associated data that VisualWorks provides to log events.

Another important factor in video documentation is lights. A camera must deal with low-contrast scenes in addition to scattered, low, and non-uniform light levels. The number of lights needed varies considerably and depends on the number of cameras fitted, site conditions, and the object being documented. Tungsten halogen lamps are simple, cheap, and used extensively under water. They offer an output that is rich in the red and orange components compared to natural daylight; this makes the light warm. It is, however, quickly cooled by the absorption of these color components in water, and at some distance it exhibits a spectral content that approximates daylight at the surface. Seawater acts as a filter, and with increasing depth the colors of solar light are gradually absorbed, starting with red, to end in total darkness. The long-wavelength red light loses about 35 percent of its intensity for each meter traveled, and the most penetrating of the shorter wavelengths, blue light, loses around 7 percent for every meter traveled.

Because of absorption it is rare that an ROV produces enough light to illuminate the large area of an archaeological site in deep water, since its power supply and lighting instrumentation are typically limited. It is therefore extremely important to try to generate as much light as possible with the least amount of power. Tungsten halogen lamps produce 18–33 lumens/W. Fully color-corrected metal vapor gas lamps are a better but much more expensive solution. These gas lamps offer up to five times more light from the same power than a similar halogen lamp; a 400 W gas lamp provides the same light as a 2000 W traditional lamp. The illumination from a gas light is also superior in many cases to more traditional lights because its focused light is better suited for close-up images. The light source of gas lamps is an electric arc across two electrodes. The arc takes place in a small, transparent capsule containing a gas or mixture of gasses, which determines the wavelength of the emitted light (Hardy, 1991).

One gas lamp, the HMI lamp (Hydrargyrum medium-arc iodide), produces 80–100 lumens/W. Whereas tungsten halogen lamps produce light with a color temperature of around 3000K, and much of this light is in the lesser penetrating red end of the spectrum, the HMI lamp produces light at around 5600K, near daylight, and much more light is present in the deeper penetrating blue and green end of the spectrum. HMI lamps are included in a class of gas arc lamps referred to as HID (high intensity discharge). HID lights have longer lamp lives than HMI lamps and are much cheaper. Used in conjunction with flood and wide flood reflectors, these lamps give remarkable wide-area illumination compared to traditional lights. However, gas lamps need heavy and bulky control gear called ballast, and in some cases this cannot be handled by a small ROV. Therefore, these lights have also been mounted on a special frame and lowered to the site to provide extra illumination.

It is also important to position the light source correctly. Ideally, light sources should be mounted at an angle to the plane of the lens of the camera. If the light source is not positioned correctly, the strong light that enters the

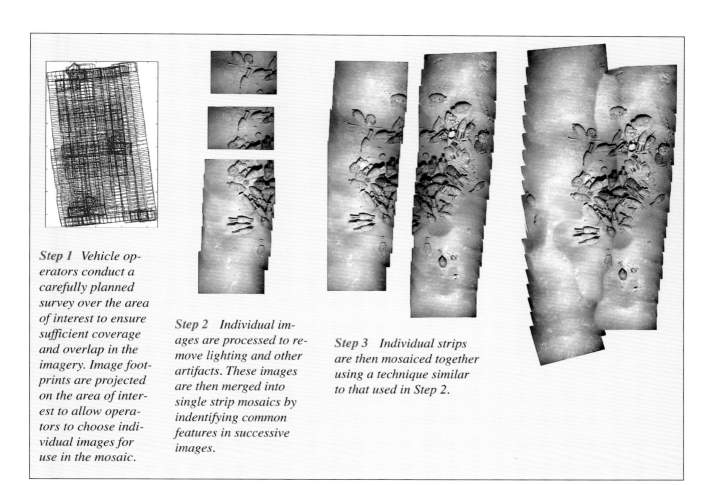

Step 1 Vehicle operators conduct a carefully planned survey over the area of interest to ensure sufficient coverage and overlap in the imagery. Image footprints are projected on the area of interest to allow operators to choose individual images for use in the mosaic.

Step 2 Individual images are processed to remove lighting and other artifacts. These images are then merged into single strip mosaics by indentifying common features in successive images.

Step 3 Individual strips are then mosaiced together using a technique similar to that used in Step 2.

The photomosaic process (Hanumant Singh, WHOI)

camera can burn out the targets altogether. This effect, called flare, occurs when a camera lens is pointed at a scene that consists of a large and quite bright area and a contrasting area that is dark or in shadow. The camera is incapable of resolving the two opposites, and all detail disappears into undifferentiated dark and bright areas.

Photomosaics

One of the fundamental problems with working under water is that light attenuates extremely rapidly and non-linearly. From a practical imaging standpoint, this means that large objects cannot be framed within a video or other optical camera's field of view. Thus, obtaining a global perspective of an archeological site requires piecing together a series of images in a process called photomosaicking. This involves running a carefully planned survey over the site, collecting a series of overlapping images, identifying common features in the overlapping imagery, and then merging the images to form strips, which are then assembled into larger mosaics (Zarzynski et al., 1995; Webster et al., 2001; Singh, Adams et al., 2000; Pizarro and Singh, 2003; Singh et al., 2004).

The major technical problem is joining the image borders so that the edge between them is not visible, by gently distorting the images near their common border so that the seam is smooth. This is usually achieved by a blending approach in which the two images are decomposed into different band-pass frequency components, merged on those levels, and then reassembled into a single seamless composite image. The idea is that with this technique the transition zone between band-pass image components can be appropriately chosen to match the scale of features in that band-pass component.

A geometric transformation must be performed on all pixels constituting each image prior to the mosaicking of the images. This must be done to bring the spatial relationships that have been distorted by the lens back

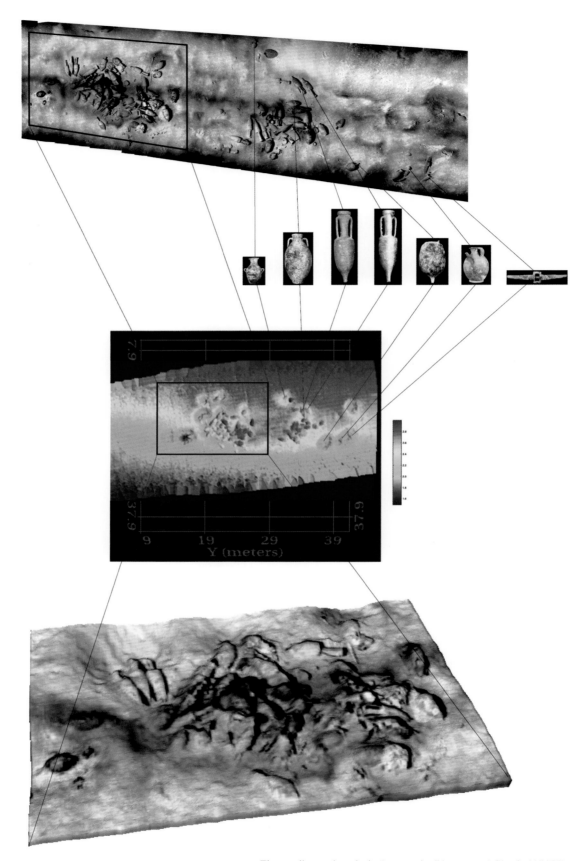

Three-dimensional photomosaic (Hanumant Singh, WHOI)

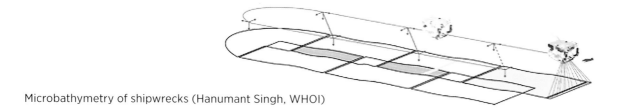

Microbathymetry of shipwrecks (Hanumant Singh, WHOI)

toward real-world values. To do this, several geometric transformational models are available. In practice, points of known coordinates are identified to the program, which then uses the chosen transformational model to warp the image so that the corresponding dimensional properties correspond to real-world values. By specifying boundaries and connecting points in overlapping images, each image is fit into the mosaic. It is possible to rotate, resize, zoom, and orient each photo. Removal of particle backscatter and color and light corrections can also be done, and a balanced light setting is important to create the impression of a seamless photomosaic. It is also possible to include an absolute or relative coordinate system. However, if the site has a considerable z-factor (vertical measurement, or depth) some of the objects will not be in exact scale relative to the seabed, unless corrections have been made using stereo photos or high-resolution bathymetric measurements. Horizontal mosaics can also be made of sites with a significant z-factor.

Photomosaics can be created by flying the ROV over an archaeological site at a constant altitude with the video or photo camera pointed parallel to the site in a true plan or elevation view. The area covered by each photo is dictated by the lens characteristic and camera height above the seabed. Using a conventional camera and flying at a height of 1 m above the seabed results in a covered seabed area of 1–2 m². Physical constraints on the distance separating cameras and lights as well as constraints on the power available for operating the lights also constitute major impediments.

The best results are obtained by flying the ROV over the site along programmed survey lines and at constant altitude when collecting the data. A human ROV operator is usually incapable of doing this with the required accuracy, but an underwater vehicle can be controlled automatically by input from motion sensors; this is often referred to as closed-loop control. In addition to an accurate acoustic underwater positioning system, control sensors include instruments that measure acceleration, attitude, depth, and altitude off the bottom. Heading is determined by instruments like a flux-gate compass and directional gyro, and acceleration is measured by accelerometers and inclinometers. These enable the ROV to be controlled by automated computer routines. The ROV can then follow prearranged tracklines while photos are taken with a vertical-mounted digital camera.

It is also possible to use high-resolution video images, captured in digital form and included in a GIS mapmaking system, as a basis for establishing a larger-scale videomosaic of an archaeological site that is both calibrated and geo-referenced.

Microbathymetry

Seabed profiling sonar, scanning sonar, and multibeam bathymetric sonar can also be used to document sites. These acoustic devices enable a profile of the seafloor to be made by stepping one or two transducers from an outward horizontal to a vertical downward position. The distance from the transducer head to the seabed or site features is recorded as a sonar return and can be used to create an image of the seabed features. By accumulating and combining the returns from the sonar pulses, models of the site can be created, such as one-dimensional hull cross sections, two-dimensional contour maps, or three-dimensional perspective views. The data from scanning sonar clearly do not have the resolution of video or photo images, but they can image three-dimensional shape and cover much larger areas than video or photo (Singh, Whitcomb et al., 2000; Pizarro et al., 2004).

The data collection process is similar to collecting data for a photomosaic. The sonar is usually mounted on an underwater vehicle and used to scan a 180-degree sector in 1- to 2-degree steps. This results in up to one hundred individual measurements for every scan, with the effective range dictated by the altitude. At 25 m above the site, the effective survey radius is 40–50 m. The data can be used to produce an x,y,z point representation of the survey area. Erroneous points must be removed and data compensated for the ROV's position and movement including pitch, roll, and yaw. The x,y,z data can then be used to form contour, three-dimensional, and section charts.

Surveying an Archaeological Site

A basic objective of surveying is to be able to determine the position of a point from some other point or points. By measuring the distance from the known point(s) to other points on the site, the other points can be plotted relative to the initial points on your plan. You can draw up a network of points joined by distance measurements on your site plan, to scale, as they are on the seabed. This is not simple, however. Sites are three-dimensional, so you must deal with differences in height or depth. Additionally, measurements always have mistakes or imprecision, and site plans get complicated when there are lots of points and measurements.

An *assessment survey* is one that aims to provide a rough idea of the extent and layout of a site; it is like a sketch with measurements. These surveys are sometimes done in advance of a predisturbance survey to provide information for planning survey work and control point positions. A good sketch often provides a large amount of information about a site in the early stages, as it is more readily interpreted than a set of bare measurements. Some form of sketch is essential before further work can be planned, since the size and shape of the site need to be known. It is important to establish the size and orientation of the site as a minimum. The position of any large features such as guns or anchors should also be recorded.

The most typical archaeological survey is a *recording survey*, including predisturbance and excavation surveys. This type of survey requires careful planning, recording, and processing.

The most simple form of positioning and taking measurements is to place a physical object on the seabed and use it as a reference. Tape, ruler, and scales have all been used to measure distances on sites with the help of a video camera. An odometer can also be used to measure distances relative to certain reference points. While it is moved between a set of points, potentiometers produce a signal that is transmitted through a cable, indicating the number of turns of the wheel and its displacement.

In shallow water the main principle is to set up control points around and inside the site. In the simplest form a diver records the position of features on a site using a distance and bearing back to a single control point. This technique is simple, gives a good enough idea of the site for an assessment, and can be done by a single diver, but it is not accurate enough for a detailed survey.

Offsets and ties can also be used to position features relative to a baseline running through the site. An offset measurement positions a feature using a single measured distance at right angles to the baseline from a known point; a tie uses two or more measurements from known points on the baseline to position the feature.

To create an accurate archaeological site plan it is necessary to record the position of points or targets in relation to some known fixed points. Any point exists in three dimensions, denoted by its x,y,z coordinates. It is possible to make a two-dimensional plan view and add the third element only when the height or depth information can increase the value of the survey. However, triangulation is the most common way to survey a site in three dimensions in shallow water. The common three-point method in which all survey measurements are made relative to datum points positioned on suitable locations around the site was developed by Norwegian divers documenting a shipwreck in the 1960s (Andersen, 1969).

The direct survey method (DSM) is the most popular triangulation solution on sites with significant *z*-factors. With at least three direct measurements from a datum point with known Cartesian coordinates, the slant range measurements meet at one point, represented by a three-dimensional coordinate (Rule, 1989, 1995). In shallow water the measurements can be done by divers with a conventional measuring tape; in deep water the measurements can be done by ROV with acoustic equipment. Measurements can be fed directly into a computer program such as Site Surveyor to make a best-fit solution for the points. This program can be used to compute the coordinates of the data points and any other structural point, artifact, or sample required. The results can be used to control all subsequent survey material in the form of conventional drawings, photos, or video. Data points can also be exported to other software packages such as AutoCad for further manipulation, reconstruction, and analysis, and a site plan can be made by mosaicking drawings, photographs, and video images over a plan of datum coordinates.

Acoustic positioning systems can effectively replace tape measures, with distances measured by sound pulses.

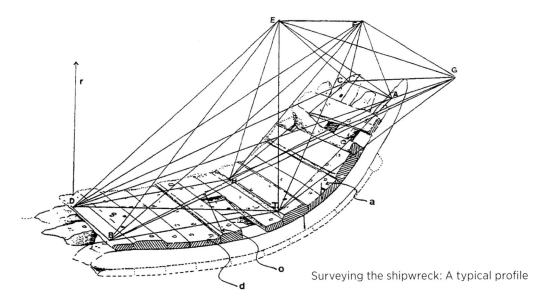

Surveying the shipwreck: A typical profile

LBL systems are probably the best acoustic solution since they offer reasonably good accuracy and are simple to use, though to achieve centimeter accuracy the data must be combined with other data and filtered. The speed of sound data must be accurate, and the equipment offsets must be known. Consider the wreck profile shown above as an example. This is a relatively complex object to measure. With an LBL system, the three-dimensional points A, B, C, D, H, and T and any other point on the structure can be found relative to the network transponders by positioning an acoustic transponder at these points. The problem with this method may be, apart from accuracy, that it requires a large number of measurements to clearly define the whole section. The position and size of single objects and their dimensions are relatively easy to establish, but it is difficult to characterize the complex shape of a timber using a LBL system, and the measurements must therefore be supported by video and photo to make acceptable drawings. SSBL systems would be used in a similar way but would not be as accurate.

Another alternative is to use an acoustic range finder. If the range finder is used systematically, it can both position and measure site features. If the reference point is E, direct measurements with an acoustic device to measure the distance between this reference point and structure points A, B, C . . . can now be made. This approach is very similar to that of a diver using a tape measure.

The accuracy of any site plan depends on the measurements. By increasing the number of measurements taken between each point and the datum points, you also increase the accuracy. Site plan accuracy should be as good as possible, but overall acceptable error varies from case to case. Research excavations typically aim for less error than ±1 cm, although the results often fall within ±10 cm. In larger surveys or during rescue operations the accuracy may be only within 100 cm. Anything larger than that and the survey is more a sketch than a plan.

Many archaeological sites on the seabed are relatively flat, so it is easy to navigate around them and relatively easy to document the site features and objects. But in the case of modern or other well-preserved sites the complexity is much higher. The complex structure of these sites can represent a hazard to navigation, since the tether can become entangled in the structure, debris, fishlines or nets, which are common on many wreck sites. Modern wrecks will soon become cultural remains, and in the United States several steel wrecks have already been investigated by archaeologists. In the Baltic Sea and Black Sea, the lack of wood-boring organisms leaves even wooden shipwrecks intact to be studied. Operations around these sites can strain the pilot, who must combine sonar and video images. In addition, there are usually several problems with using acoustic equipment at these sites, including acoustic shadowing and multipath errors. In general, the complex three-dimensional structure of these sites is also more difficult to document.

Photogrammetry

Photogrammetry, which involves the measurement of horizontal and vertical geometry from photographs, can be used to measure and position artifacts and site features

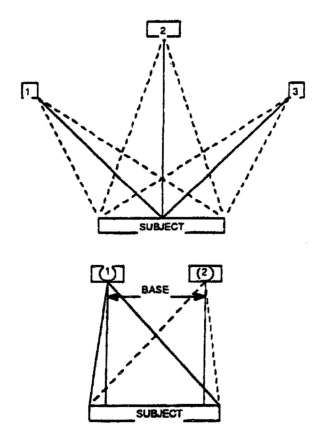

Principles of photogrammetry (Garrison)

under water (Garrison, 1992). By analyzing the information contained in two stereo photographs of the same scene, taken with a calibrated camera from different angles, measurements in all three dimensions can be obtained directly from the single pair of photographs with good accuracy. Photogrammetry is based on the principle of colinearity, which posits a direct linear relationship between points in a photographic image and the same points in the world or object space. Points in the two worlds can be related under the assumption that the camera lens provides perfect central projection and that the focal plane of the camera is perfectly flat. Since no camera lens system is perfect, small errors have to be accepted or variations from lens distortion, focal length, and radial distortion corrected for. As a rule, the smaller the focal length, the wider the field of view. Photographs are taken with a camera that incorporates a grid of accurately spaced reference marks on the film platen. Each negative then incorporates a highly accurate scale and avoids distortions created by film variations or environmental effects.

To obtain a measurement in three dimensions, more than one image is required, and it is usually common to use two cameras. If the object distance/base distance ratio is more than 7:1, accuracy falls rapidly. The object distance is the distance between camera and target object, and the base distance is that between the two cameras. It is also possible to use only one camera, in which case two or more images are taken at converging angles and there are several unambiguous points or objects common to all of the overlapping images. Unless a digital still camera is used, photos must be digitized. Image measuring and processing algorithms are used to convert image coordinates into three-dimensional object coordinates. The algorithm determines the exact position of a camera when each photograph was made. Thus, the camera operator need not maintain a strict base distance separation, although results are usually best when photos are taken at roughly right angles to each other.

To give distances and measurements obtained by photogrammetry an absolute true scale, it is necessary to have a known object in the image. Image calibration can be achieved by placing a reference object such as a metric rod or cube in the image. Small low-powered green or red lasers can also be used for underwater scaling and measurements (Tusting, 1996). Instead of having an object with known dimensions in the image, which must be moved around with the ROV, it is easier to use two lasers with known separation. The laser dots will be visible in the images and can be used for scaling.

The typical precision of a photogrammetry system configured for medium-range inspection tasks is in the region of 0.5 mm for an object of around 5 m, after calibration of the system and camera lens. Each object must be given a unique identification number, and three-dimensional coordinates of each object in the image can be transferred, for instance, to a CAD program to produce two- or three-dimensional site plans. A complete site plan can be made by mosaicking the plan of datum coordinates found from several sets of photos covering the complete site. It is also possible to create planimetric photomosaics based on stereo photos.

Photogrammetry has the advantages of being a remote sensing technique that can be used to assess objects of almost any size and orientation. It is also suited for ROV operations. Precision can be adjusted by changing the geometry of the photos, in particular the separation of the cameras, the distance to the object, and the focal length of the cameras. The photos can also be easily reassessed. Photogrammetry is a relatively cheap solution but has nonetheless seen limited use in underwater archaeology—mainly because of poor image quality and the fact that, since light attenuates extremely rapidly and nonlinearly, it is difficult to image large objects underwater with a camera.

There are, however, some examples of photogrammetry in underwater archaeology, and the method has been used to produce detailed record drawings of shipwrecks. In the 1960s, a shipwreck site covering 6 by 12 m was recorded with two 70 mm cameras. A site plan was then created with an accuracy of ± 40 mm (plan). In Cyprus, a similar map at a scale of 1:10 was made of a fourth-century shipwreck covering an area of 4 by 8 m using two Nikonos still cameras. A control frame was placed over the site to allow onsite calibration. Here, an accuracy of ±10 mm in plan and ±25 mm in height was achieved. In Canada, a map was created of a Basque ship, wrecked in 1565, using two Nikonos cameras. This map had a planimetric accuracy of ±10 mm (Newton, 1989).

Photogrammetry enables object positioning and measurements in three dimensions. In the wreck illustrated above, stereo photos of the wreck section could be taken from at least two of the positions E, F, or G. Using a calibrated system and a reference object in the wreck section, the positions of points A, B, C, . . . could then be accurately defined and distances in all planes calculated.

On discontinuous sites with large distances between site features, positioning must be done by an acoustic method since mosaicking scaled photos leads to large errors. But individual objects can be measured with photogrammetry, and sections of the site can also be documented with a photo-based method. On a continuous site, photogrammetry is better suited to document details than acoustic methods.

Another simple solution is to work only in a two-dimensional plan projection. Two-dimensional data can be obtained from any single photograph that satisfies the following conditions: the data or object plane and the film plane are parallel, and the scale is known. Then any measurement of the film image multiplied by the ratio between the image scale and the object scale is valid (excluding measurement errors and lens distortion of video and photo cameras). If, for instance a video camera is placed at a distance r from the site in the shipwreck section above, distances on deck timber d can be found if a scaled object has been placed in the same plane. Deck timber a is in a different image plane, and an attempt to measure it from the same point would lead to errors. If the scaled object is placed in the image plane created by timber a, and the video camera is turned until the image is parallel to that plane, deck timber a can be measured and positioned.

To make a complete image, the two results must be combined. Scaling can be achieved by paired laser dots or physical objects. Alternatively, a laser-generated line or stripe can be projected into the video camera's field of view. In addition to planar distance measurements, laser lines can also facilitate accurate measurements of convex or concave surface features, since the line will show a visible offset over three-dimensional features. Since accurate measurements can be made only in the image plane, measurements and positioning of advanced three-dimensional features are difficult unless the object is recorded from many angles.

PhotoModeler is a well-known Windows program that transforms photos into three-dimensional models. To create a model with PhotoModeler, several photos must be taken of an object from different directions but from roughly the same distance. The photographs are imported into the PhotoModeler system and displayed on the screen. The operator then marks features of interest by tracing over the photos with the mouse. All points must be cross-referenced between the different photos. When the photos have been marked, PhotoModeler makes a three-dimensional model from the marked photos. The marks on the photos become points, lines, or surfaces in a single three-dimensional space. A known distance must be assigned to the model so that it becomes scaled. After a model has been produced, accurate measurements can be taken from it and models can be exported to three-dimensional CAD programs.

The accuracy of PhotoModeler depends on the resolution of the camera and the size of the objects being measured. A high-resolution image can ideally give a precision of 1/3000 of the object extent, and lower-resolution images can give a precision of 1/300 of the object extent. The precision is also influenced by the camera calibration and the care taken by the operator when marking the photos. Because PhotoModeler uses a camera as a measuring device, it needs information about how a particular camera bends the light rays. Therefore the camera focal length, image aspect ratio, image position, and lens distortion must be calculated. This is achieved by taking photos of a supplied pattern and importing these into a calibration program. PhotoModeler represent a cost-effective solution but is at its best when modeling objects with well-defined visual features, which is not always common on a marine archaeological site (Ewins and Pilgrim, 1997).

Defining the Buried Content of a Shipwreck

Archaeological material is often completely buried beneath sediments and therefore cannot be located by visual aids alone. If a site is covered by sediments, it is possible to define the buried content with the aid of a sub-bottom profiling system.

The CHIRP system has been used extensively for this purpose in Denmark and England (Quinn et al., 1996). A CHIRP system transmits a computer-generated, calibrated FM pulse that is linearly swept over a full-spectrum frequency range, also called a CHIRP pulse—for example, 2–16 kHz over 20 ms, although the range can be anywhere from 400 Hz to 24 kHz. This wide choice of frequencies and selectable CHIRP bandwidths allow the operator to optimize the system configuration for sediment penetration and layer/object resolution, and a vertical sub-bottom layer and object resolution of 6–8 cm can be achieved. The acoustic returns to the system are matched and filtered with the outgoing FM pulse, generating a high-resolution image of the sub-bottom stratigraphy.

The resolution is entirely dependent on pulse length. High-frequency sources produce a high-resolution image of the bottom without much penetration. A low-frequency pulse penetrates deep but produces low-detail images. A conventional system operating around 3–12 kHz has limited penetration and a vertical resolution of around 20 cm. Seismic boomers have a high-energy output that offers deep penetration but a resolution which is typically limited to 1 m. The CHIRP systems that use wide-bandwidth transducers limit this trade-off between penetration and vertical resolution by transmitting a range of frequencies.

The University of Southampton tested a CHIRP system to determine whether it could image wooden artifacts buried in shallow marine sediments (Quinn et al., 1997b). Since wood is not a homogenous material (e.g., growth irregularities from tree to tree), the reflection coefficients obtained from wood vary considerably. A series of laboratory experiments were carried out on an oak half beam taken from the wreck of the *Mary Rose* and compared with some theoretically derived values of oak. The beam values were found to be lower than those predicted by theory. This may be explained by the large variations in elastic properties between individual wood specimens. The outer segments of the waterlogged wood were degraded and showed more homogenous elastic properties, whereas the heartwood had not degraded. Traditionally, wooden wrecks are composed of oak, with lesser components of mahogany, pine, and elm.

Since the reflection coefficients of wood and sediment are considerably different, oak buried in marine sediments should be readily imaged by a suitable sub-bottom profiling system. Similar results were achieved for ten other wood species. This knowledge may be used in the future to process and interpret sub-bottom data better. Knowing the typical reflection coefficients from a variety of materials, it is possible to estimate whether an anomaly is likely to be cultural or a natural feature and thereby decrease the number of possible targets. Some calculations of reflection coefficients acquired over the *Invincible*, an eighteenth-century buried oak wreck, support these results (Quinn et al., 1998).

The high resolution offered by CHIRP indicates that it can be used to search for buried marine archaeological sites. It was found that artifacts must be larger than 1 m in the horizontal plane and greater than 15 cm in the vertical plane to be detected in depths of less than 30 m using a typical configuration. Still, the two major problems related to searching for a site using a sub-bottom profiler are not solved by the CHIRP system. Being a single-beam system, it must be towed along closely separated search lines, and obviously the survey speed will be very slow. In addition, even though this research represents the first attempt to distinguish reflection patterns of cultural remains and natural features, the problem has by no means been solved, and thus sub-bottom profilers are probably better suited for surveying known sites.

This was, for instance, done on the *Mary Rose* site. A survey was carried out to investigate the site for remaining wreck structure and to assess the applicability of the CHIRP system to marine archaeology. When the collected data were processed, two new large anomalies close to the original wreck site were discovered. These anomalies were erosional scour features created by the wrecked obstacle lying on, and partially buried within, the seabed. These results indicated that the in-filled scour features could contain important, preserved material. More detailed inspection by divers confirmed this, and several additional pieces and artifacts from the *Mary Rose* have now been recovered (Holt et al., 2004), clearly showing that sub-bottom profilers, rather than merely being used to locate sites, can be used to define the extent of a site more accurately.

Topographic parametric sonar (TOPAS) is another high-performance seabed and sub-seabed inspection system (Dybedal et al., 1986). The advantage of TOPAS is the parametric source, which creates a high-directivity, low-frequency beam from a very small transducer. Parametric sonar arrays utilize the nonlinear propagation properties of water to generate a low-frequency signal in the water column based on two high frequency or primary signals (Webb, 1993). The low-frequency, secondary signal is generated at the difference between the two high-frequency, high-intensity signals. This technique makes it possible to generate a narrow-beam, low-frequency signal that is very stable and repetitive with a moderate-size transducer. The secondary signal wavelets can be Ricker wavelets, narrow-band signals, CHIRP wavelets, or phase-coded wavelets. The resulting TOPAS signal is around 5 kHz with a beam width of approximately 4–5 degrees. Most CHIRP systems have a beam width of approximately 20 degrees. This means that TOPAS has much better lateral resolution than most other systems. This is particularly important in searches for complex structures. The TOPAS beam can also be steered electronically to perform sequential beam scans, generating sub-bottom swaths and building up three-dimensional displays of topographic and seismic profiling data. This is the only sub-bottom profiling multibeam system available today with a 40-degree scan sector to each side. The TOPAS system has been used by NTNU to define the buried content of several shipwreck sites.

MIT is also working on a high-frequency, narrow-beam, sub-bottom profiler to "see" down into the mud. Managed by computerized control and mapping, the resulting data should allow archaeologists to record and replay a "virtual excavation" of a wreck site, that is, a three-dimensional model, removable in layers. This device has a much higher frequency (150 kHz) and a very narrow beam (3–5 degrees). Although it does not penetrate nearly as deep into the mud as its lower-frequency cousins, the narrow beam allows the instrument to make detailed images that can depict small, buried features.

The sub-bottom analysis of the Ormen Lange site was one of the most comprehensive sub-bottom surveys ever conducted in deep water over a shipwreck site (Bryn et al., 2007). Three different sub-bottom data sets were collected over the site and in the planned pipeline corridors (see chapter 3). The sub-bottom data sets were collected with a Geochirp II system in CODA format, Innomar data in SES format, and Edgetech FSSB data in SEG-Y format. All data was reviewed in native format to ensure that there was no degradation in resolution. Targets were selected only if they were single point-sources or a contiguous series of anomalies or diffractions, regardless of amplitude, in the near surface sediment and not deeper than the preglacial-postglacial sediment boundary. Features or anomalies that were obviously of geological origin, with or without seafloor expression, were not selected as targets. Anomalies that were the result of acoustic, electrical, or environmental noise in the data were also excluded from selection as targets whenever possible. The data were plotted in ArcGIS over the available sidescan sonar mosaic and correlated with cultural material and modern debris to determine the origin of the anomaly in its respective data set.

Three areas of the site were selected for excavation on the basis of the distribution of clusters of sub-bottom profile anomalies. These areas were found in relatively close proximity to the wreck and along the pipeline routes, with priority given to anomalies lying along the pipeline corridor and between the corridor and the wreck. In this way, anomalies or artifacts at risk of disturbance or damage because of their proximity to the pipeline installation, as well as the corridor for construction activity, would be excavated.

Within the site of the extant wreck there was an apparent correlation between the higher density of localized higher-amplitude reflectors and disturbed sub-seabed sediments associated with the location of the wreck. The buried content near the main wreck site was postulated to extend several meters to the west and southwest by the sub-bottom profile data, and this was subsequently confirmed by excavation. In the pipeline corridors farther away there was no such evidence of a direct relationship between the presence of a higher-amplitude reflector in the data and any buried artifact. Most anomalies were, rather, indicative of soil attributes (high organic content, shell fragments, and clastic material), showing once again that it is difficult to distinguish cultural remains with sub bottom profilers.

Magnetometers can also be used to chart remains even when these are completely buried. Experience has proved that within a certain level of intensity most artifacts on a site, including most nonmetallic objects, are marked out within a magnetic anomaly. A magnetometer can therefore also be using during site assessment, since it allows the area containing debris from a site to be circumscribed.

On the Ormen Lange project, geomagnetic data were collected at the shipwreck site and throughout proposed pipeline routes with a Geometrics G822 marine magnetometer affixed to the survey ROV. The magnetometer data confirmed the main results from the sub-bottom analysis, but the data were not detailed enough to identify small sections of the wreck or individual artifacts outside the main wreck area. However, the general trend of the migration of artifacts and site taphonomy substantiated the data provided by the sub-bottom profilers.

Acoustic Cameras

Acoustic lenses and cameras use a large number of acoustic transducers placed on a grid and are particularly suited for viewing three-dimensional underwater scenes and objects. Several acoustic cameras have been produced. One such system has two crossed arrays, each contains 64 elementary transducers that send and receive beams. The system provides a three-dimensional image based on the 64 focused beams operating at 500 kHz (Hansen and Andersen, 1996). The image can be viewed on a computer screen. Using three-dimensional scene reconstruction algorithms and volume visualization techniques, a scene and objects can be recreated, although the resolution is low compared to photography. The acoustic images can, however, also be used to establish positions and dimensions of objects in the image and can be used in low-visibility conditions.

The Echoscope is another acoustic system that delivers high-resolution three-dimensional images under water in real time. Originally developed by OmniTech of Norway, now part of CodaOctopus, it employs a highly innovative multielement transducer design, high-speed digital signal processing, advanced sonar beam formation techniques, and intuitive image display software to create three-dimensional sonar images. Ensonifying the whole viewing volume with a single ping, Echoscope uses phased array technology to generate over 16,000 beams (128 by 128 beam array) simultaneously, resulting in a three-dimensional sonar image. With up to twenty updates per second, the echoscope can be used for real-time monitoring and has several advantages over multibeam echosounders since it can acquire all the data required for an inspection task from a single ping, removing the need to coregister the data with survey-positioning and motion-sensing equipment.

Laser Line Scan Cameras

Laser line scan systems minimize the effects of backscatter and can therefore be used to create images that cover large areas of a site. With a laser line scan system it is possible to create photo-quality site plans with centimeter accuracy from 10–20 m above the site. Large-scale mosaics are also possible, and the laser imaging systems reduce the need for ordinary photomosaics. The availability and cost of these systems must, however, be improved before they are used extensively in archaeology (Debrule et al., 1995).

Taking Samples

In addition to video and photo work to position and measure artifacts and main site features, the documentation phase often includes selected sampling and recovery of important artifacts. Artifacts are collected to confirm the age and identity of the site, and wood samples are often collected for dating. The most common way to collect these samples is to use a manipulator arm, mounted on an ROV system. The sample can then be brought to the surface in the arm or placed in a collecting basket. Coring is also sometimes used to assess the buried contents of a site and can be achieved by mounting an electric or hydraulic corer on the ROV or by pushing the corer into the sediment with the help of the ROV's thruster power. For an elaboration of sampling tools and techniques, see chapter 6.

Storing Archaeological Data

Archaeological fieldwork may be viewed as a research task that undergoes phases with increasing focus and detail, as seen in the next figure (Søreide et al., 1996; Søreide, 1999). The initial planning phase is characterized by an effort to identify potential sites of interest and involves an accumulation of all the historical transcripts and other relevant written material concerning settlements, industry, and trade. This may lead to potential archaeological sites where both the location and the degree of remaining material are highly uncertain. Relevant collected data can be stored in a database as textual documents with illustrations. Prior to the search, these data must be analyzed to identify the most promising areas related to finding deepwater sites. The results can typically be represented in a database as maps and geographic coordinates.

The objective of the search phase is to confirm the existence of the archaeological site and is undertaken as a systematic survey using several types of sensors, including sidescan sonar, sub-bottom profilers, and magnetometers, depending on the actual conditions. The resulting images can be stored in a database and further analyzed in the computer to improve the results. Since the basis for planning and executing a survey is a map of the area, there is obvious potential here for application of GIS technology. The map can be used as a reference for indicating promising areas, planning search tracks, and pinpointing interesting finds. The map may also be supplemented with more advanced bathymetric surveys and echo-sounder profiles.

If the search phase identifies interesting archaeological sites on the seafloor, the remaining phases concentrate on more detailed investigations through close-up observation and excavation followed by conservation and analysis. Site documentation and excavation result in an abundance of data from the sensors concerning positioning and measurements of artifacts, descriptions, drawings, photos, site plans, and more. The lowest level in the data hierarchy is object representation, which may come from several sources: close-up video and photo, drawings, reference to samples and supporting text.

An archaeological project thus accumulates an increasing amount of information about the site, the seafloor conditions, object locations, and individual objects. As the flowchart indicates, there is a basic relationship between project phases and data segments. In practice, however, any aspect can be addressed at any stage during a project. Additionally, data segments are hierarchically related. For instance, an artifact with certain characteristics is located in a specific place on the seafloor; the overall geography and topology of the seafloor are therefore primary characteristics of the study area.

An important aspect of marine archaeological work is that tasks often have to be done within strict budget and time constraints. It is, therefore, important that the

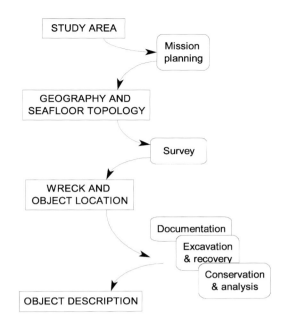

Information hierarchy and processing model

capture and storage of data be efficient, and that the data are highly accessible for later study and analysis. This is a most challenging objective given the fact that much of the data originates from various systems or media, as shown in the figure below.

A database is usually defined as a collection of related data and generally deals with pieces of standardized information. In complex applications like an archaeological database system, the database must be capable of handling large amounts of data with many different data formats and types. The database must therefore be an object-oriented multimedia database that can handle complex objects and operations. A multimedia database must store data from several different media sources, including numeric data, texts, bitmap images, raster graphics images, video, and so forth.

At the lowest level, a database is a collection of data related to a particular project, including aspects of a single site or artifacts, although a specific database may also be part of a more general-purpose database, such as computerized museum records. Because of the computer capacity needed and problems associated with defining standard formats for large general databases, project-specific databases are the most common.

A project-specific database can have several formats. A database created in the first phases of a marine archaeological project has different content and thus format than

STUDY AREA
- Transcripts of historical documents
- Trade and marine traffic patterns
- Analysis of human activity
- Hypothetical locations of sites
- Survey plan (mission)

GEOGRAPHY AND SEAFLOOR TOPOLOGY
- Charts, geographic information systems (GIS)
- 3D - perspective drawings
- Echo sounder profiles, bathymetric surveys
- Sea floor material sampling

WRECK AND OBJECT LOCATION
- Trackline of executed surveys
- Applied survey vehicles and equipment
- Log, position & characterization of finds
- Drawing of location (lay-out), distances
- Supporting documentation: Sonar plots, echo sounder plots, sub-bottom profiles

OBJECT DESCRIPTION
- Close up video and photos of objects
- Measurements (optical/acoustic)
- Computer-aided-drawings (CAD)
- Data / findings from samples
- Object evaluation: Type, origin, age, use etc.

Data sources in marine archaeology

a database created in the final two phases. The result of the search phase is usually a site inventory, with coordinates and short descriptions of the located sites given in a GIS (maps, coordinates), typically backed by survey information such as sidescan images.

It is critical for an efficient system that the subdatabases are linked and can be compared. A database management system is a collection of programs that enables users to create and maintain a database. Designed correctly, the database management system can also represent complex relationships among the data. The archaeological data must also be described in a standardized manner in a database, so that they can be meaningfully compared. Since many data consist of observations about artifacts and their contexts, an object can be represented as a data element with attributes. The attributes describe the element and have several different states. Attributes can be dimensions, material type, and so forth. It is also possible to include contextual attributes such as the environment or association with other objects. Above the artifact level is the group of artifacts, often called assemblages (Richards and Ryan, 1985).

When collecting archaeological data, one must identify and record the site elements and their attributes as economically and completely as possible (Steffy, 1994). In theory, every artifact can have an almost infinite number of attributes, but in practice only the necessary attributes according to the level of analysis chosen in the research design should be collected. An archaeologist who is interested in the complete site will clearly use other attributes than one interested only in the pottery. The choice of what data to record, how to measure, and the manipulative techniques to employ remain subjective decisions. If possible, however, an attribute definition should follow an established convention. By using international standards, such as functional classifications of ships and artifacts or soil color charts, objectivity is improved (Cederlund, 1988). Quantification of the data also makes it easier to manipulate the data set, but there are clearly problems related to quantifying subjective factors (Hill, 1994).

That said, the practice of archaeological classification implies more than merely ordering the finds of an excavation, and even a classification reflects interpretations of the data. Depending on the research questions, a classification may have many different levels of abstraction. In the case of a shipwreck, it is common to distinguish using the general function of an artifact and its specific use on board the ship. One possible classification model consists of six main categories (Gawronski, 1992): *Ship, Cargo, Armament, Equipment, Personal possessions,* and *Environment.* These six categories can be further subdivided into categories of artifacts that fit within a certain functional context. Unfortunately, concepts such as *Cargo* can be used with different meanings, and it is important that the classification chosen represents a find in a coherent way. A classification other than function may therefore prove more appropriate, such as *Parts of ship, Artifacts, Parts of artifacts,* and *Nonartifactual remains.*

A typical video inspection produces hours of video as well as handwritten observer logs and printouts from associated instruments and sensors. By consolidating the observer logs, instrument readings, and video images into the database, the data logging process is clearly made more efficient. Digital video can be viewed in a window on a computer screen. The inspectors can log information on the same screen, and important video frames can be stored in the database together with additional data. Reports can be distributed on CD or DVD or over the Internet in near real time (MacDonald and Juniper, 1997).

Processing Archaeological Data

After the investigation of an archaeological site has been finished, archaeologists can get all the necessary data for the postprocessing phase from the database. Data stored in the database can be used for processing and modeling in a classical archaeological analysis and for hypothesis formation, evaluation, and publication of information. This processing work can be seen as the manipulation of the collected data to produce additional data in a more useful form, usually called information. Whereas data processing in the field must be supported by structured data acquisition, a more multimedia- and document-centered approach based on the collected data in a distributed, multiuser database is necessary in the postprocessing phase. The full information processing system must therefore include a database level (one or several databases, as described above), a processing level with several different software applications, and an information management level.

A database can be used to store and access the data, but additional computer programs are needed to analyze and process the collected archaeological data to create new information in the postprocessing phase. Several existing and purpose-made software applications can be used to process the data including, statistics, modeling, graphics, word processing, and several others. This can be illustrated by the following example. One thousand artifacts have been found during an investigation. Their positions are represented as position coordinates in a database. These coordinates can be used in the postprocessing phase for statistical purposes to establish quantities of certain objects, major locations of certain objects, and the like. But these data points can also be used in a graphic software package to create site plans, or as the basis for a written analysis using a word processing package. These three software applications are therefore capable of creating new information based on the same data set. The outputs of one processing iteration may be the input of another processing cycle, since the actual processing can involve any of the following (Richards and Ryan, 1985):

- duplication: reproducing the data in identical or edited form
- verification: checking the data for errors
- classification: separating the data into various categories
- sorting: arranging the data in a specified order
- merging: taking two or more sets of data and putting them together to form a single set
- calculation: performing numerical operations
- selection: extracting specific data items from a set of data

The purpose of a data and information management level is to administer and track all data processing to avoid duplication of effort, increase efficiency, and link all interdependent information. In the field and in the postprocessing phase, several different members of the team usually need to access, control, and share parts of the collected data, and this is difficult to manage. Thus it is necessary to use an information management system to integrate many of the information processing tasks in a way that leads to reliability, quality, and utility of the data and supports both the individual and collaborative aspects of the process. An information management system therefore enables the team members to share the relevant data.

The processing of the collected data results in the creation of documents. An information or document management system offers more sophisticated storage, retrieval, and control of these documents. Using an archive system, it is easy to locate, view, and edit any document, and related documents can be linked. For example, a drawing in AutoCad, a photo, and a text file can be linked to define an artifact completely. This merging and fusion of the project information is a significant step forward over conventional manual methods and less advanced computer-based solutions.

Document management systems can store, access, and handle a diversity of information, such as CAD drawings, text, spreadsheets, databases, and sonar images and make it simple to organize, locate, retrieve, share, and store documents. A document management system uses a revision control system to make sure team members always work on the latest version. These revisions are visible to the other users. Documents must be checked out and checked in from a system computer vault. By using so-called viewers, it is not necessary to open the software programs to see the content of a file. The project manager can issue a work order by sending requests with attached documents to other scientists, and the files can then be added or edited according to user status and returned to the project manager for review. Thus the document management system enables team members to share and access relevant information in one system, except actual artifacts. Various representations of artifacts such as drawings, photos, and textual descriptions are, however, usually sufficient for most purposes.

In theory, the information processing system should be custom-built for marine archaeology, but it is often more cost effective to use a combination of tailor-made and common software. The price of common software is lower, but these systems are usually less suited than a tailor-made solution to solve the task. It is important to always consider the possibilities for integration between the various systems when creating an information system, using some common systems and some tailor-made solutions. A system's objectives usually change over time, and new tasks have to be added to the system. It is therefore likely that the system will take on more and more tailor-made features with each modification. It is also important to remember that, although data systems contain data, an information system must also consider the human operators and their interplay. The computer system is useful only to the extent that it is properly integrated in its environment and must be planned in a total context, taking all relevant factors into account including the technical systems, the software systems, and the human organization.

The importance of an information processing system cannot, however, be underestimated. Documenting underwater archaeological research in satisfactory fashion is difficult, time consuming, and expensive, but the standards of documentation are important, because these are some of the most important sources for making conclusions and hypotheses. A successful information processing system is thus important to ensure that the information is used in the best possible manner.

Summary

The best documentation results are often achieved by combining the results from several types of sensors, so-called sensor fusion (Stewart, 1991). Minerals Management Services general procedures for documentation of shipwrecks can be used to summarize how to document sites in deep water. These recommend that a professional archaeologist be on board the survey vessel during the investigation of unidentified magnetic anomalies, sidescan sonar targets, and shipwrecks, or at least that a professional archaeologist review the findings:

1. During ROV investigations of magnetometer anomalies, sidescan sonar targets, or shipwrecks, the ROV pilot is prohibited from disturbing or picking up any artifacts, including features and other structural components of a shipwreck.

2. Do not turn off the video feed any time during the survey. Keep a complete record, from the time you approach the site to the time you leave it.

3. During ROV investigations, the ROV pilot must not allow the tether to drag on the site or in the debris field.

4. Use an acoustic underwater positioning system. Fixes (waypoints) are to be taken near the location of the object (if visible) to provide information on the size and shape of the anomaly.

5. If the anomaly is a shipwreck, the fixes should be taken on the bow, amidships port-side, amidships starboard-side, stern, and any other survey ties that would help determine the size of the vessel. Include important structural components such as stacks, pilot house, external machinery, housings, and artifacts.

6. In addition to the physical remains of the shipwreck, survey the surrounding area around the vessel to determine the size of the debris field. This may include the area between the proposed area of impact and the wreck site. Determine position of and video-survey any significant debris. Use scanning sonar from as many positions as needed to define size and geometry of the debris field.

7. It may be necessary to run a series of transect lines to get the entire site.

8. Request the surveyor to take timed-sequence x,y,z points so that it is possible to reproduce a post-plot of ROV movement.

9. Complete video inspection with voiceover commentary during all survey activities. Continuous video with position overlay is to be recorded throughout the entire survey. Be systematic so that complete coverage of the sides of the vessel is recorded smoothly in sequence. In addition, record features such as stacks, pilot house, damage, external machinery, and housings. Do not get distracted by interesting marine life. It is important to be able to follow where you are on the wreck, which can be difficult if you go offline to chase a fish.

10. Video as closely as possible without disturbing the site all unique features such as superstructure, stacks, pilot house, damage, external machinery, housings, debris, and artifacts and continuously record all on video tape. For all unique features, record a video TIF image (clip).

11. Thoroughly inspect bow and stern for the name of a vessel.

12. If possible, position the ROV on at least four points around the vessel at a sufficient distance from the wreck to image the entire vessel on the ROV scanning sonar. At each of these points, make both plan (sector scan) and profile imagery of the wreck using the dual cursor measuring system to define the size of the wreck. At each of these points, record TIFs of each sonar measurement in addition to continuous digital or video tape.

13. Note and survey any significant scars on the seafloor in the immediate vicinity of the wreck. Collect video and sonar imagery of seafloor features such as drag scars, mounding, or depressions associated with the wreck. In addition, take footage of any industry-related items in the immediate vicinity (e.g., lost anchors, boreholes, pipelines, discarded pipe).

This methodology will provide the most information on the site under investigation. If recorded anomalies cannot be located using the above search methods, additional ROV transect lines may be required to determine the precise location and limits of an anomaly or anomaly cluster.

SIX

Excavation of Deepwater Sites

If all of the questions concerning an archaeological site cannot be answered by the information collected in the documentation phase, or if the site is particularly interesting, it is sometimes necessary to investigate it further and carry out an excavation. Since the cost of an excavation phase is high, excavations are typically done only on sites that represent an important archaeological contribution or sites that will be totally destroyed, for example, by construction work on the seabed. Complete excavations are seldom carried out, because of the cost and complexity.

Alternative Intervention Platforms

The major problem to solve when excavating underwater sites with ROVs is that the systems weigh hundreds or thousands of kilograms and risk disturbing or even destroying the very sites they are sent to investigate. Additionally, accurate and delicate maneuvering of an ROV on an archaeological site is a challenge. This is why most deepwater archaeological operations have been limited to documentation, sampling, and digging trial trenches.

An archaeological excavation phase involves several advanced tasks, like the careful removal of sediments to recover artifacts, mapping and picking up artifacts, documenting the site features and the excavation process, and raising artifacts to the surface. These tasks challenge the capability of any remote intervention system, especially advanced manipulative tasks. The best-suited remote intervention system for the excavation phase is the work-class ROV system, which can carry the tools needed for excavation and advanced documentation tasks. The size of the ROV and tools depends on the depth and complexity of the operation. In moderate depths and with a limited scope of work, a small work-class ROV may be sufficient, whereas for really deep or complex projects a large work-class ROV or scientific ROV is necessary. Alternatively, it is possible to use a manned submersible or bottom-crawling ROV, which moves primarily by exerting traction forces on the seabed via a wheel or track system, as long as this does not destroy the site.

Most archaeological sites under water cover only a relatively small area of the seafloor. Continuous sites require different investigation methods than discontinuous sites. On a continuous site, one approach is to attach the ROV to a frame. A frame set over an archaeological site prevents the robot from accidentally coming into contact with fragile artifacts and allows archaeologists to maneuver the ROV with more precision.

The ROV docks onto the frame so that it sits only a few centimeters above the wreck site on four adjustable legs. The docking platform can move in all directions on the frame by motorized cogwheels. Positioning is based on rotation sensors on the frame, backed up by high-resolution directional sonar sensors or acoustic positioning systems, with resulting subcentimeter accuracy. The frame therefore provides unique positioning control, rivaling that of a land excavation.

Alternatively, the frame and ROV can be replaced by so-called remotely operated tools (ROTs) like those developed by the oil and gas industry to do advanced tasks on the seabed. An ROT system contains the tools needed to complete the tasks on the seabed and is task specific. The ROT can be lowered to the site on the seabed with a winch and positioned by thrusters or guidewires. To position the ROT on the correct location in deep water, it is

ROV excavation of a shipwreck site in combination with a seabed frame. (Brynjar Wiig)

necessary to use a dynamically positioned or anchored surface support vessel. When the ROT has been successfully installed on the worksite, it can be used to complete the specified work. Usually the ROT is controlled by a separate umbilical cable from a topside control station.

The idea of using ROTs for deepwater archaeological investigation instead of ROVs is interesting. The ROT can complete the same tasks as an ROV, perhaps even better since it has been purpose-built to do the necessary tasks. The technical complexity and cost could also be lower in some cases. ROT systems can take many shapes and solutions. By far the simplest solution would be to mount the necessary equipment such as video cameras, lights, manipulator arms, and pumps on a static metal frame. This frame could then be lowered to the site with a winch and positioned in the work area, as seen in chapter 3. At a wreck site in Norway it was possible to position a static frame at 70 m and to complete documentation work using a video camera and take samples using a manipulator arm. Such an arrangement can also be adapted so that it is possible to move the arm and camera on the frame (rotate/lateral movements) or to introduce excavation equipment and collector tanks.

A more advanced version would include possibilities for the ROT to cover a larger area of the site in one operation. This would either mean a system of booms that could move the equipment around or a system in which the equipment position on the frame can be shifted, perhaps by actuators or cogwheels. The ROT could, for instance, be designed so that it consists of a frame that covers the complete site or parts of the site. A robot could then move around on the frame to do the necessary tasks, including documentation and excavation. The biggest problem with an ROT solution is positioning it on the seafloor, which requires a substantial surface vessel to maneuver and install the heavy and potentially large structure.

It is also difficult to design ROTs that are suitable for more than one type of archaeological site. A ROT should therefore be modular so that it can be adjusted to suit a particular site. Before a ROT is designed, the terrain and site characteristics must be documented; then the optimal ROT configuration can be designed and adjusted to suit the site and terrain. A ROT solution is obviously best suited for sites that are fairly continuous and with limited vertical variation (i.e., spread out in two dimensions only). If the site is discontinuous, the size of the area favors an ROV operation. If the site is well preserved with three-dimensional structures, the ROT solution is also less effective than an ROV operation, since the main advantage of the ROV is its flexibility and ability to move around.

Other special solutions to install the necessary equipment on an archaeological site without disturbing it have also been proposed. One consists of a boom crane that carries the equipment. This can be either a stand-alone system or ROV-mounted.

Complete excavation of a site is a tedious operation, as can be demonstrated by the well-preserved *Wasa* site

(Franzen, 1966). The hull of this ship consisted of around 14,000 large and small pieces, thousands of which had broken off and lay on the seabed, where divers had to document and retrieve them. In addition, some 9,000 artifacts were found in and around the ship itself. Most shipwrecks contain thousands of artifacts. Investigating, excavating, recording, and conserving archaeological sites is clearly a complex task.

Removing Sediments

There are two phases in an underwater archaeological excavation. First the area to be excavated must be uncovered. Typically there is a lot of sediment covering a site, and this must be removed to reveal the uppermost layer of the site features, including construction details and artifacts. The second phase includes more detailed excavation of individual items and is usually done layer by layer. To remove the sediments covering the uppermost layers of an archaeological site, it is possible to use a thruster. This method has been used on several sites. The sediments are blown away with a thruster mounted on the ROV or ROT. The thruster flow must be adjusted to control the speed of the process. By using two thrusters instead of one, a well-designed system can be nonrecoiling and can be operated by a free-swimming ROV. With a single thruster it is necessary to dock an ROV on the seafloor during the operation.

An alternative to a conventional thruster is the jet prop system—a four-bladed propeller housed in a duct. The propeller is driven by seawater. Using a pump, the water exits through four tangential pressure jets, one at the tip of each propeller blade. Thus in effect the jet prop system converts a low-volume, high-velocity flow into a higher-volume, low-velocity flow. This low-velocity flow impinges on the seabed soil, stirring it up into a suspension, and creates an outward flow that carries the sediments with it and excavates the seafloor. Fine-tuning is achieved by adjusting the system's operational height above the seabed, or by using a larger or smaller jet prop system. The jet prop system is typically ROV-mounted. It is also possible to excavate in hard sediments with a low- or high-pressure water jet system.

Another alternative is to use a dredge to excavate. Suction dredging is the most popular excavation method for other seabed excavation purposes. Most dredges use an eductor pump. The power for the eductor section of the pump is provided by a stream of water driven by a centrifugal impeller. The water is drawn into the inlet and passes through the impeller section of the power pump before being injected into the main suction stream in the annular eductor nozzle. This high-velocity fluid creates a low-pressure region behind the suction nozzle that provides the power to excavate.

When removing the first layers of sediments covering an archaeological site, it is important to monitor and calculate the depth, width, and rate of the excavation in order to avoid damage to site features. This can be achieved in real time with sonar and video cameras. It is, however, difficult to get good images with cameras because of the mud in suspension. Scanning or bathymetric sonar is a better solution and can create an extremely accurate longitudinal profile of the excavation area, which can be used to ensure that just enough soil is removed to uncover the buried features of the site.

When the uppermost layer of sediment has been removed and the site features are revealed, it is necessary to excavate in more detail around each object. An excavation typically involves documentation and removal of layer by layer. An excavation can be completed in many different ways. A full archaeological excavation can, for example, be done either by excavating a site in one large operation, or area by area. In many cases, however, only small sections of the complete site are excavated, as with a trial trench. Wherever it is possible, the stratigraphy surrounding the artifact must be systematically excavated and recorded until the item is totally free of the sediment. It is then possible to remove the object. An object should never be pulled from the sediment, which risks damaging it and failing to record associations to nearby objects.

The more detailed excavations can be done by a variety of tools. A manipulator arm with a small shovel or brush can be used to remove sediment. Alternatively, some sort of dredge is common. The Ormen Lange project utilized an ejector dredge developed especially for marine archaeology. This ROV dredge consisted of a water pump, an ejector, and a suction hose with a suction head. The ROV operated the suction head with the manipulator. The water pump was powered through a hydraulic interface on the ROV, so no extra power supply was required. The suction head was a failsafe in the sense that it was impossible to block. The suction head was also equipped with water jetting to disintegrate clay and other cohesive material.

Ormen Lange suction dredge handled by manipulator arm (NTNU Vitenskapsmuseet)

Investigating the contents of the collection tank at Ormen Lange (NTNU Vitenskapsmuseet)

Suction picker on the Ormen Lange site; items were stored in a collection device (NTNU Vitenskapsmuseet)

Two dredgers were constructed, one 2-inch and one 3-inch. The suction force could be adjusted to fit actual conditions. The suction head was made from clear plastic, enabling monitoring of the water flow. Artifacts pulled through the system were filtered through a sediment collection box on the seafloor, which was connected via a flexible hose to the exhaust of the excavation dredge. The box content was later checked for small fragments. It is, however, important to note that ideally an excavation system should not be used to collect objects on an archaeological site. This must be done by other devices, such as manipulator arms that can pick up artifacts without unnecessary damage.

Other projects have also used a combination of dredges and water jets. Odyssey Marine Exploration uses a dredge to excavate wreck sites and developed a complex filter system to pull objects out of the silt before they are ejected out the exhaust and into receiver tanks that retain solid material and extrude the seawater. The manipulator arm delivers the excavation nozzle to the excavation site. The system and receiver tanks are ROV-mounted, and the contents are inspected on recovery of the ROV each day. On the SS *Central America* operation (see chapter 3), the ROV was also equipped with a hydraulic dustpan that swept up smaller artifacts with a jet of water.

Another popular excavation tool is the airlift, which consists of a rigid tube. When compressed air is injected at the lower end of the tube, usually by a compressor at the surface, the air rises toward the surface through the tube, creating a suction effect at the bottom. Water and loose material near the tube opening are pulled in and up. The power is dependent on the difference in depth-related pressure between the top and the bottom and the amount of air injected. The effect can be controlled with a valve. An airlift is therefore similar to a water dredge, differing only in that the suction effect is created by air flow instead of high-velocity water flow. An airlift is better suited for diver operations in shallow water than ROV operations in deep water because of the long tube lengths required at depth.

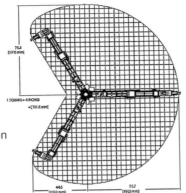 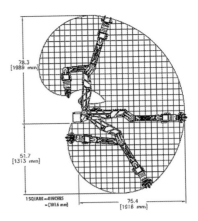

Typical range of motion for a large seven-function manipulator arm. Left: top view, Right: side view (Schilling Robotics).

Picking Up Small Artifacts

To pick up artifacts, a remote intervention system must be fitted out with robotic arms, commonly known as manipulators. Small items may be removed by an arm simply by taking hold of it with the claws. Manipulator arms have from one to seven functions; having more functions enables the arm to hold additional positions in the operating envelope. The functions are configured after the human arm:

1. Claw open/close
2. Claw rotate
3. Wrist pivot (left/right)
4. Forearm rotate
5. Elbow pivot (left right)
6. Shoulder up/down
7. Shoulder slew (left/right)

Some arms also include extensions (left/right/rotate) as additional functions.

Manipulator arms are referred to by the number of functions they provide. Many of the one- to five-function arms are electrically powered. As the tasks become more complex, stronger and more functional arms are required, and larger manipulator arms are therefore hydraulically powered. Each function is controlled by a dedicated hydraulic valve. The motion of an arm on the seabed is controlled by a master arm on the surface, and the slave manipulator on the seabed duplicates every movement of the master arm. The operators can do advanced manipulative tasks by moving the master arm and studying video images of the slave arm. The master arm is normally a kinematic replica of the slave arm, with the same relative range of motion. The position-controlled closed-loop servo system allows movement introduced by the master arm to be duplicated by the slave arm. An ROV and manipulator arm operation is often called teleoperation, since the operation is viewed and controlled by an operator.

The best solution when using a manipulator arm to pick up objects during an excavation is to park the ROV on the seafloor or on a frame above the seafloor. This reduces the degrees of freedom of the operation from six to three, since the complete pick-up operation can be done by the manipulator arm alone. To be able to pick up an object, the manipulator arm must have at least four or five functions. If the arm has fewer functions, the ROV must be moved around to compensate for the lack. This is, however, much more difficult. A dual-arm configuration is perhaps the most flexible solution, since two arms can complement and support each other during advanced operations. It is also possible to use booms to increase the work envelope of a manipulator arm, and stereo cameras to increase the depth perception of the operator.

Force feedback and bilateral servo systems provide an electronic means to control the pick-up operation with the greatest dexterity and without excess or accidental force, since the operator can feel the load as it is applied. The master control has small motor-driven actuators on each joint that provide the operator with a feeling of the load of the slave arm. The force feedback manipulator systems can be used to pick up even the most delicate artifacts (Macy, 1993).

In addition to the typical claw, several tools or so-called end tooling can be interfaced with an arm and used to pick up objects. When designing tools for lifting archaeological artifacts, it is important to examine the artifacts. On a shipwreck site many construction items are large and made from wood. These require special solutions. Other artifacts vary in size, from small buttons to anchors and cannon, although the majority are in the range of 5–30 cm. Typical archaeological objects include the following (Robinson, 1981):

> Wood: timber and wood substances such as cork or bark
> Other organic materials: plant products (textiles, matting, rope, paper) and animal products (bone, ivory, horn, wool, leather)
> Inorganic materials: glass, pottery, stone
> Ferrous metals and concretions
> Copper and its alloys: bronze, brass
> Other metals: lead, pewter, silver

Artifacts under water have already been affected by several deteriorating factors, which need to be considered when you are planning the pick-up operation. Most wooden objects have typically been attacked by wood-boring organisms and are waterlogged, with a soft and spongy surface. It is therefore important to be careful when lifting a wooden or other fragile artifact. Many objects can be lifted with a manipulator arm, but the claw must not have sharp metallic edges that may damage the surface of the softened wood. Larger pieces of wood and timbers require special lifting devices and additional support to avoid damaging the wood, which may not be able to support its own weight when lifted.

Appropriate lifting techniques utilize supporting plates, frames, and covers as well as cushioning and additional support made from, for example, wet foam, plaster of paris, epoxy resins, or polyurethane. Many items can also be wrapped and padded in plastic boxes, crates lined with foam rubber, or water-filled polythene bags. Alternatively, fragile objects can be raised in large sections, surrounded with their burial sediments—but this approach may very well cause damage to the surrounding stratigraphy and should be used with care.

Organic material is an important marine archaeological discovery, since it is rarely discovered, but sometimes organic material can be preserved under water, especially if covered by sediments. When uncovered, most organic materials deteriorate rapidly and are extremely delicate, so they should be handled as little as possible. After having been fully recorded in situ, the material should be packed in plastic containers while still on the seafloor, sandwiched between layers of polythene, and covered

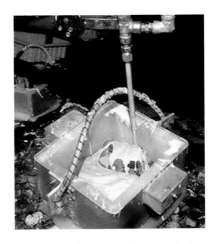

A silicone injection system used at the SS *Central America* wreck site to lift several objects in one operation without damage (Milt Butterworth, SS *Central America* Discovery Group)

by sediment. It can then be taken safely to the surface. When a fragile object is exposed, it is important to protect it temporarily from events such as abrasion, sediment movement, and marine fauna. Ideally, such items are picked up and brought to the surface immediately, since the cycle of deterioration starts as soon as the item has been exposed. But it should be remembered that bringing artifacts to the surface also represents a major environmental change. It is therefore important to have a team of conservation specialists present to take care of recovered material, and a decision to raise objects must be taken only if the subsequent conservation can be guaranteed.

Other specialized lifting tools can be used for particular objects. The best of these are probably suction pickers, which can lift all sorts of artifacts including, wood, ceramics, coins, and other metal objects. The Ormen Lange project utilized a small suction cup connected to a pump by a hose to pick up more than two hundred artifacts of various types. Several other deepwater archaeology projects have also used suction pickers with success. Artifacts can also be lifted by specially developed tooling constructed on-site. A silicone injection system, for instance, has been developed to lift several objects in one operation, without damage. Special end-effecter assemblies were designed to lift Roman amphoras by Ballard's team in the Mediterranean. These were designed to cradle the amphoras softly in soft, synthetic fish netting. Other tools can be designed for additional artifacts, usually in combination with the manipulator arm.

When a stained glass window from the *Titanic* was found lying intact on the seafloor, a lifting support was constructed and taken down to the shipwreck site. The manipulator arm carefully slipped the baseboard under the window frame and placed a cover in position on the top. The window and lifting support were then brought to the surface in a lifting basket (Hutchinson, 1994). These examples and many other projects have demonstrated that even the most delicate objects can be recovered.

Raising Artifacts to the Surface

The manipulator arm and the other lifting tools can pick up only one artifact at a time. It is possible to carry each object separately to the surface in the manipulator claw, but if there are numerous artifacts this becomes a tedious and time-consuming operation. This approach also gives the artifact poor protection, especially in the splash zone at the surface. Another and better solution is to collect several artifacts in a collection device and raise this to the surface when it is full. Thus the ROV does not have to surface every time an object is collected. There are two main systems to do this: artifacts can be collected in a storage device on the ROV or in a completely separate collection device.

Collection devices carried by the ROV include several types of external containers and small baskets. It is common to add extra buoyancy material to reduce the in-water weight of these devices so that they can be handled efficiently by the ROV. It is also possible to have special compartments in the ROV to store items. These can be drawers that come out of the ROV or bin samplers, a set of rotating boxes placed on a skid below the ROV. A manipulator arm can then place the collected artifacts in the boxes. Ideally the storage containers are of several different sizes and be covered in protective foam for safe transport to the surface. The lifting capacity of an ROV is totally dependent on the thrust capacity of its vertical thruster(s), since ROVs are typically only slightly buoyant in water. This places a practical limit on the number of artifacts an ROV can carry.

Another alternative is the separate storage system, a container used to store and raise artifacts. The container can be positioned on a site with an underwater positioning system by winch or ROV. When the compartments in the container are filled up, the container can be hauled to the surface. Another solution is to use an autonomous elevator device, which free-falls to the seafloor. The ROV can use its scanning sonar to find the elevator device and use its manipulator claw to transfer an artifact to the elevator compartment. Since the distance between the site and the elevator can be great, this solution is not suited for major excavations with several thousand artifacts. The elevator can be ordered acoustically to rise to the surface by releasing a weight. It is also possible to use a guide wire system to position a storage container on a site.

A collection device must be designed to protect an artifact when it passes through the splash zone. If artifacts have not been satisfactorily secured, the movements of the device and the wave and water action can damage them or even wash them out of the container. It is best if the container can be closed.

Raising Large Artifacts

A small item can be picked up simply by taking hold of it with the manipulator arm, but the lifting capacity of the largest manipulator arms is limited to about 90 kg, so it is often necessary to attach hooks, slings, nets, strings, or plate grippers to a large object and lift it by cable from the surface ship or with lifting balloons. Objects that cannot be lifted by manipulator arm include cannon, anchors, and large parts of vessel. A wooden structure is seldom raised to the surface because of the high cost of conservation and because it can instead be documented in situ and reburied. Whenever a structure is considered important enough to be raised to the surface, it is usually necessary to dismantle parts of the structure and lift it in smaller sections. Pulling apart the structure simply to make artifact recovery easier is totally unjustifiable. Before dismantling, it is necessary to record the timber structure completely. Alternatively, casts of the structure can be made with rubber compounds or other suitable substances. Constant recording is also necessary during the disassembly.

After detailed inspection has been undertaken, the amount of preparation required to lift the different artifacts must also be determined. For instance, wooden structures must be sound enough to withstand the lift forces and not to break up in the lifting process. Thus, it is important to determine the best possible attachment points for the lifting device and to design additional support needed for both dismantling and lifting. The objects must be strengthened or supported, especially where the

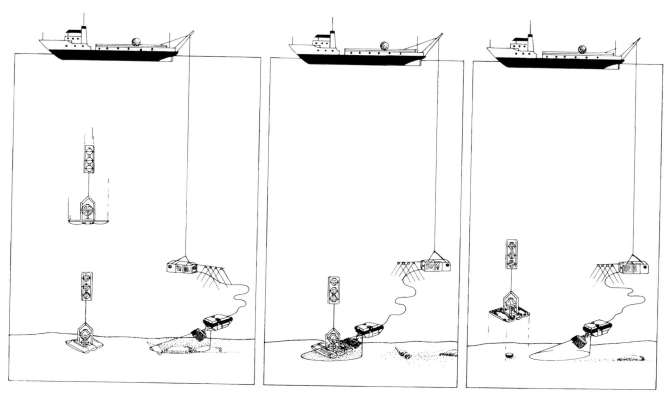
An autonomous elevator device used by the Jason team

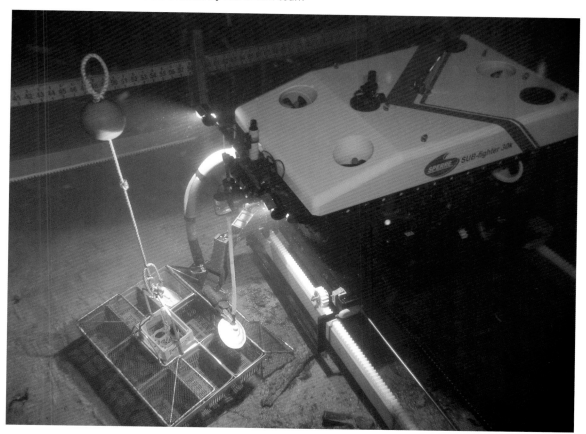
Elevator from the Ormen Lange project (NTNU Vitenskapsmuseet)

lifting load is applied, for only limited deformations can be allowed. The force needed to break an object out of the sediments should be minimized by dredging and increasing the flow of water through the soil.

One of the most popular methods of lifting large objects is to lower a lift line from a surface ship to the seabed. The lift line should have a weight some distance from the lower end, which keeps it taut and near vertical even in strong currents. Scanning sonar mounted on an ROV can be used to find the weight in the water column. When the weight has been located, the ROV can move in and catch the lower end of the lift line. While holding the line in its manipulator arm, the ROV can maneuver back to the site, pulling the lift line behind. The line can then be attached to an already prepared attachment point on the object. The lift operation can be observed through ROV-mounted cameras. After attachment, the lift line tension is increased and the object hauled up.

Lift line operations are simple in theory, but in deep water it is difficult to position the lift line and necessary to use a heave-compensated winch to avoid the motion of the ship jerking and destroying the object. An alternative to lifting the actual object is attaching the object to a frame or transferring it to a box or a crate, and lifting this instead.

It is also possible to use a salvage claw to pick up objects that are too heavy or large for a manipulator arm. A salvage claw can either be mounted under an ROV or operated by a manipulator arm. The ROV places the claw around the object so that it can be picked up. Again, though, the ROV usually has neutral buoyancy in water, so its lifting capacity is limited by the power of its vertical thrusters. In special cases a variable ballast system that uses compressed oil as a seawater displacement medium has increased the lifting capacity of an ROV. In most cases, though, the better solution is to use a lift line from the salvage tool to the surface as described above.

Another alternative is to use a lifting balloon to raise larger objects from the site. Air is by far the most common buoyancy medium under water. It is cheap, and its low density gives it a large lifting force per unit volume. The air balloon can be attached to an attachment point on the object and the lifting process is started by pumping air into the balloon, until the object is free and raising. It is, however, difficult to use an air balloon in a deepwater environment because of the high pressure. Because of the compressibility of air, its lifting power decreases with increasing water depth; to fill an air balloon at 2,000 m requires air pressure of at least 200 bar. In addition, an air balloon is unstable, since the air expands with the reduction of pressure; without a buoyancy control system, it rises faster and faster without control (Snowball, 1996).

An alternative is to use fresh water in a balloon (Stangroom, 1995). The density of fresh water is roughly 2 percent less than that of seawater. This is enough to give substantial lift in the balloon, although a fresh-water balloon clearly needs a much larger volume to lift an object than an air balloon. Lifting one ton would require a balloon with roughly 50 m^3 of fresh water, compared to 1 m^3 of air. The water balloon can, however, be filled from the surface quite easily. The column of fresh water gives rise to a hydrodynamic pressure almost identical to that of the pressure of seawater, and the pump therefore has to overcome only the 2 percent difference of absolute pressure, equal to 4 bar at 2,000 m. On the other hand, the long supply cables are clearly a disadvantage in deeper water. The lifting process can be controlled by opening and closing valves, and a fresh-water balloon usually has a sensitivity of less than 10 kg, so it can be used to increase lift forces very slowly and move in both directions. The lift balloon can also be filled with diesel, which is slightly less dense than seawater, but this represents an environmental hazard.

An alternative water balloon system consists of a thin-walled, spherical pressure vessel that contains a fluid (usually seawater) and a hydraulic pump to pump the fluid out of the vessel. As the fluid is pumped from the vessel, the space created above the fluid creates an upthrust that corresponds to the amount of fluid removed from the vessel. Clearly the thickness of the pressure vessel must be adequate to withstand the external pressure (Clyens and Automarine, 1996). Thus, the cost of such a device increases progressively with depth.

Raising Complete Sites to the Surface

The risk and cost of deepwater archaeological investigations point to an alternative solution in which the entire archaeological site is moved to a more favorable location for excavation. It is possible, at least in theory, to lift a site intact from the bottom and place it on a specially designed water-filled barge or in a tank to be excavated. A tank could also have viewing windows, which might provide both a more controlled environment and public access that could generate additional revenues.

The extremely well-preserved, seventeenth-century Swedish navy vessel *Wasa* was found in 1956, and the Swedish government financed a massive rescue operation in which a team of shore-based archaeologists and engineers raised the complete wooden warship. Navy divers worked in zero visibility conditions and used a recoilless water jet to make six tunnels through the clay beneath the ship. Each tunnel was 1 m high and 3 m wide. An airlift sucked the mud to the surface, where archaeologists sifted for artifacts. When the tunnel work was complete, the divers passed cables through the tunnels and attached them to two big semisubmerged pontoons above. These pontoons were slowly raised by filling them with air, and thus the *Wasa* was gently lifted off the bottom and step by step eased into shallow water. The main part of the shipwreck was excavated in a museum built to house the ship while divers systematically excavated the rest of the site.

The British warship *Mary Rose* capsized and sank in the Solent in July 1545. After a preliminary study of the remains, it was revealed that many of the personal possessions of the crew and the store were preserved in situ and that the surviving hull was a coherent structure that might be salvageable. The decision to salvage the hull of this historic ship was, however, not an easy one; consideration had to be given to the economic feasibility of the project and the long-term restoration and conservation program required after the salvage. The site was finally excavated in four main diving seasons in 1979 to 1982 and revealed a cross section of artifacts from the Tudor ship. The remaining hull was an open shell, consisting mainly of the starboard side of the hull, rather than a complete cross section with transverse strength (Dobbs, 1995).

The lifting operation consisted of several steps. The hull was first wired to a lifting frame. The system of lifting wires was designed so that the loads of the lifting process could be more evenly distributed over the entire wood structure. Rather than dig tunnels under the ship and pass strops through them, some sixty-seven main lifting points were positioned and holes drilled in the structure. A bolt was then placed in each lifting point, and using waterjets and airlift divers dug under the hull to locate the other end of the point, where a backing plate was fitted and the bolt tightened up. Each wire was made to the specific length between the point and the lifting

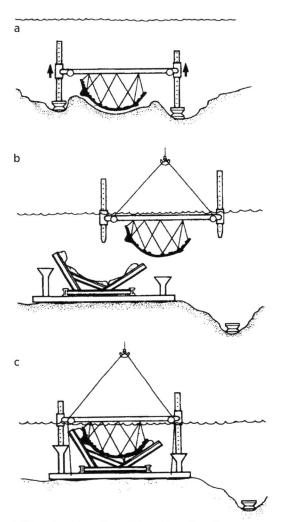

Lifting the *Mary Rose* (*Mary Rose* Trust)

frame, and equal tension was achieved with bottle screws on each wire.

Hydraulic jacks operating on the legs of the lifting frame were then used to raise the hull the first critical centimeters until it was free of the suction effect of the silt below. The hull, hanging from the lifting frame, could then be transferred underwater to a lifting cradle. Precise positioning was required to steer the legs of the frame into the four guideposts on the cradle. The cradle had been built to conform to the shape of the hull, using section drawings produced in the archaeological survey. When the hull had been safely placed in the cradle, the entire structure was lifted into air by a floating crane. The *Mary Rose* is now on display in Portsmouth. The entire lift operation cost US$5 million excluding gifts and conservation.

In 2003 a new program began on the *Mary Rose* site. A deeper and wider channel to Portsmouth harbor required the excavation of as much as possible of the two spoil heaps flanking either side of the hole left by the recovered hull. These were formed during the original excavation by the removal of hull sediments through unfiltered airlifts on opposing tidal runs. A tracked ROV system was custom-built specifically for excavation, burial, and trenching work. The ROV was equipped with an airlift suction head to recover all sediments to the surface for screening. This was the first use of remote excavation on a historical wreck site in the United Kingdom. A Sonardyne LBL system was used to position both the ROV and divers who assisted in the operation. This method enabled the team to recover over 380 objects in three weeks, a wide range of items including arrows, coins, and dagger handles.

On February 17, 1864, the CSS *H. L. Hunley* became the first submarine to attack and sink a ship during a war. After the attack on the USS *Housatonic*, the *Hunley* was lost with all hands. The submarine was recovered 136 years later by a team of engineers and divers using a support truss capable of lifting the entire vessel intact. Along with the lifting truss, the engineers designed a sling/foam system for supporting the submarine during the lift—a critical step because of the delicate condition of the *Hunley*'s hull. A tank for storing the *Hunley* at the conservation lab in Charleston, South Carolina, was also designed and built to ensure the integrity of the hull.

Wasa and *Mary Rose* could be easily accessed by divers, resting in less than 30 m of water. The task of raising complete and continuous archaeological sites is even more complex and expensive in deep water. The approach used on the *Wasa* and *Mary Rose* can be applied but would require advanced ROTs and ROV operations. Another alternative is large claws that surround the site and sediments and lift the complete site to the surface (Bascom, 1976). It has also been suggested that a site or large section of a site could be covered in silicone or liquid nitrogen, then lifted in one operation. Covered in liquid nitrogen, an artifact in a cluster would have a constant temperature of around -3°C. Lifting could be achieved with a balloon or lift line as described in the previous section, and conventional archaeological investigations could then be carried out in a laboratory, probably using less time and at a lower cost than ordinary underwater investigations. The liquid nitrogen method could also save organic remains, which are often damaged in conventional excavations. In general, more work remains before suitable methods for lifting complete sites from deep water can be tested.

Documentation

Documentation is a vital part of any excavation phase. For each excavation layer, the uncovered area and artifacts must be documented by video and photos and positioned and measured. The excavation process itself should also be recorded. It is important to record all relevant information.

An object that is uncovered must be given a unique identification number and positioned and measured using acoustic methods or photogrammetry. The geographic reference points used for the positioning should ideally be identical to those used in the predisturbance plan, although it is now more important to record vertical measurements, which can become substantial in excavation trenches. The position of the trench, context, stratigraphy, as well as environmental evidence should also be recorded, and a short description and interpretation of each artifact, associations, and coordination with other material should also be included. An online data system to record relevant information collected during an excavation phase is needed.

In deep water, the reference point for the collected data should be time, and each record must therefore be time-stamped. Time can link video and still images to identification numbers of artifacts and related information such as position and size. Data in the computer system can then easily be processed and used by archaeologists in the postprocessing phase. Artifacts should be removed from the collection device as soon as it is on deck, tagged, labeled, and entered in a log. They should also be measured and photographed as soon as possible, allowing first for appropriate conservation procedures.

When it is discovered, wood is often fragile or soft and prone to damage from tools. Care should therefore be taken to avoid the risk of surface scoring by tools. If complete timbers are raised, it is essential to record the precise location, orientation, angle, and inclination of all elements, so that their original positions can later be accessed. The appearance and natural features of wooden objects, such as surface features, size, and condition, should also be recorded. Since many surface features are

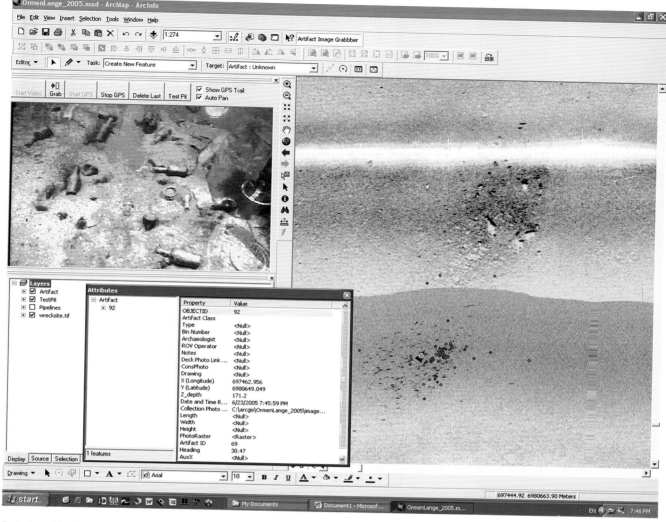

Data input to the ESRI system (NTNU Vitenskapsmuseet)

most visible while the wood is wet, it is important to complete this work in situ. Associations with other elements on an archaeological site must also be recorded, to provide functional and technological interpretations of the elements.

An alternative positioning system can be developed for a frame-based excavation, such as that used on the Ormen Lange project. When a positioning device on the frame has been moved into position, the objects underneath can be positioned precisely and given a x,y coordinates relative to the frame. We can return to the wreck sketched in chapter 5 for an example. The frame-based excavation tooling is positioned over the site, and the site is excavated until the profile shown in the figure is uncovered. Points A, B, C, . . . can be positioned (x,y) by moving the positioning device over each point and using a video camera with a crosshair sight or laser as a pointing device. The z coordinate can be found using an altimeter and would be equal to r for point D.

Recording Systems for Deepwater Excavations

During excavation of a deepwater site it is necessary to record information in an organized manner. The information from an excavation phase typically concerns both content and context, that is, what is found where in the sediment layers. When sediments are removed, artifacts and parts of the structure are revealed. Every item must be well documented using video and photo cameras, and detailed measurements and positioning of each item must also be carried out (Gifford, 1993).

Site Recorder, operator (right) and screen (below) (Peter Holt)

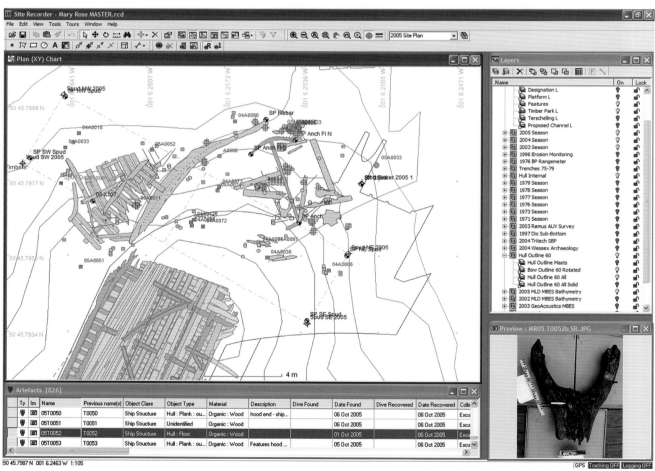

Several proprietary recording systems have been developed for deepwater archaeological projects. The Ormen Lange project contracted with ESRI to develop a software module in ArcMap to aid in recording images and data about artifacts as they were excavated from the shipwreck. ArcGIS is an integrated collection of GIS software products for building a complete GIS system designed for storing, manipulating, analyzing, and displaying data in a geographic context. The module consisted of a dockable window in ArcMap that can be used to show a live video stream from the ROV in concert with an analogue-to-digital capture device and Microsoft's DirectShow API. It can also parse custom positioning feeds input from multiple locator devices that provided x,y,z position inputs for the ROV and x,y positions of artifacts relative to the ROV and update the map display. Users can click a button that

- Inputs the x,y position of an artifact from the positioning sensors on the subsea excavation frame.
- Captures a still image from the video stream and writes it to disk.
- Creates a point feature representing the artifact in a geodatabase feature class at the current position.
- Completes attribution information about artifacts (type of artifact, material, etc.) using relevant subtypes and coded value domains.
- Stores the image information as a raster attribute so that it can be previewed on various backgrounds (sidescan, site plans, multibeam, etc.) within ArcMap.

Video from the ROV-mounted cameras was stored digitally using VisualSoft digital video recording systems. Various additional attributes were stored with the video in VisualSoft, which enables users to coordinate data in the ESRI database with digital video.

Site Recorder, from 3H Consulting, is another versatile GIS system specifically designed for marine archaeological projects. It replaces the separate surveying, drawing, handling of finds, and reporting programs usually used on a site. Separate pieces of information can be associated with one another, allowing easy analysis and interpretation. The program integrates geographic charts with information databases and survey processing tools to allow a site to be documented completely. Unlike most other GIS programs, Site Recorder is designed for collecting information and not just displaying it, and it can be used for all phases of archaeological work:

Search. Site Recorder can assist the search phase of an operation by collecting and displaying all information about a site. Magnetometer, sidescan sonar, and multibeam sonar targets can be imported and plotted on a site plan along with information about each target.

Assessment. Site Recorder can be used during assessment work on a site by collecting and processing survey information. Site Recorder includes drawing tools similar to those found in CAD programs, so a detailed sketch can be quickly drawn up showing the important features of the site. Unlike other GIS programs, the survey points can be used as a framework for the drawing.

Predisturbance recording. Site Recorder supports full three-dimensional survey data collection and processing using distance, depth, and position measurements. The program uses a survey-quality least-squares adjustment to compute the optimum position of survey points and hence the finds positioned relative to them. It supports three-dimensional trilateration, offset and ties (baseline trilateration), and radial surveys.

Excavation recording. For recording during excavation, Site Recorder becomes a program for handling finds, including information about artifacts, contexts, and samples. Detailed information about these objects can be recorded in the program as well as being positioned "live" on the site plan, A standard set of properties has been defined for each of the object types with user-configurable word lists for the appropriate properties.

Site Recorder can take in live position data from GPS or acoustic underwater positioning systems and use it to show ships and ROVs moving around the site plan. Position fixes can be recorded at any time and used to position finds and survey points.

Site management and monitoring. As well as finds lists, survey data, and site plans, Site Recorder can record dive logs and a list of people associated with the site. The information is stored in a hierarchy of levels used to represent set phases of work on the site. A season's work can be gathered together in a project, and this may contain many layers of different types of related information. Data collected during surveys or site visits can also be

(Right) Documenting artifacts (NTNU Vitenskapsmuseet)

Field conservation on the Ormen Lange project (NTNU Vitenskapsmuseet)

grouped in this way. Map and chart drawings can be collected together in base maps, again using multiple layers for each.

All of the information about the site is readily available in reports or files exported in common nonproprietary formats. A data management system like Site Recorder make postprocessing of collected data much easier.

Postprocessing

During fieldwork, evidence is collected in a variety of forms, and a wide range of techniques must be applied to extract the maximum information from it. The collected artifacts have to go through conservation and further documentation in the laboratories, while a team of archaeologists and other specialists process drawings and photo records to generate site plans and mosaics, acquire specialist opinion, and analyze and interpret the collected data and supporting evidence from several additional sources, such as archives. The purpose of this work is to write the archaeological report.

SEVEN

Preservation Conditions in Deep Water

The contributions of marine archaeology have typically involved access to sources not available to land-based archaeologists. After the introduction of scuba diving equipment, it became possible to pursue questions that land-based archaeologists had been forced to ignore. New and so far little-known sources from the seabed, like ships, cargo, and submerged constructions, were becoming available to give new and improved understandings of the maritime context.

The contributions from deepwater sites can be significant. It is possible that wreck sites in deep water include unknown items or information to supplement sources found on land or in shallow water. Unknown types of ships and cargo can give new data about trade patterns and trade routes, crafts, levels of technology, the development of ship building, political structure, and more. These contributions can benefit studies on trade, naval warfare, and historical chronologies.

Internationally, marine archaeology has been biased toward shipwreck sites. In deep water, where wrecks and objects lost or thrown from ships constitute the majority of the finds, this bias is even stronger. But finds from other kinds of sites, such as prehistoric settlements, would also be significant. For example, preserved organic remains from prehistoric sites would be very interesting. Unfortunately, the high cost of deepwater archaeology has meant that spectacular shipwrecks, treasure wrecks, depth records, and in some cases shipwrecks threatened by destruction have been the focus.

Even though there are fewer remains in deep water than along the coast, they may be of higher archaeological value if they have been better protected against destruction, wear, and man-made interference or been better preserved. Under special conditions, preservation of organic materials, such as food remains, is better in water. Deepwater archaeology therefore represents a chance to find materials that are not present in shallow water or on land because of different preservation conditions or formation processes.

It has been argued that deepwater conditions in some cases lead to perfect preservation, and several stories exist about perfectly preserved ships on the seabed (Tolson, 2005). But even though such sites would be of particular importance and value, deepwater archaeology should not be biased toward only finding things particularly old or well preserved. The clue needed to answer a specific question may just as well be found in less-preserved sites. Certain object types, like ceramics, glass, and some metals, will always have survived and can give important information. Archaeology in deep water should have the same goals and principles as marine archaeology in shallow water, though based on different methods.

It has also been argued that deepwater shipwrecks are archaeologically more complete than their shallow-water counterparts and thus worth the added cost of their investigation. But this case has yet to be proved through scientifically collected data.

Site Formation Processes in Deep Water

Site formation and site classification models have become common features of wreck studies as tools for explaining the formation of a site and the effects formation processes have had on the archaeological evidence contained in the site. Site formation processes transform a sinking vessel into the site as observed from the archaeologist's point of view. We can study these processes to find out how well preserved deepwater sites may be and, in particular, whether deepwater sites are better preserved than those in shallow water. If it is true that the preservation conditions are more favorable in deep water, this will clearly increase the deepwater potential.

The Process of Wrecking

Various estimates have been made about how many ships have been lost in the open sea well away from the coast and in deep water. The majority of shipwrecks are caused by some accident in connection with the shore, but several estimates indicate that 10–20 percent of all ships sink in deep water. The most important factors for ships sinking in open waters are

- Weather conditions. A ship can be overwhelmed by storm winds and waves; battered by these forces the ship could be blown over, the hull run leak, the mast could break and damage the hull, ballast stones or cargo could shift position and ruin stability, or the ship could plow into the backside of a wave and disappear.
- Mechanical/physical damage not caused by weather, such as poor design, corrosion, rot.
- Collision with other ship, ice, or other obstacle.
- Fire or explosion.
- Act of war (sunk by the enemy).
- Running aground but sinking in deep water because of currents or steep underwater terrain.
- Deliberate abandonment, such as when a ship becomes useless, is junked by pirates who want only the cargo, or is scuttled for military purposes.

The process of wrecking is typically different in deep water than in shallow water and raises a variety of questions: Do the ships break up, do they sink differently, do they penetrate the seabed? A hull will probably remain more intact in deep water than in shallow water because it avoids the major physical damage caused by wave action when battered against a shore line, but in some cases (for instance, explosion) the hull will have suffered major damage. If a wooden ship breaks up on the surface, its parts and contents are likely to be spread over a large area, since the greater part of an old ship is made of material that usually floats unless it becomes waterlogged.

When a relatively intact ship sinks in deep water, it can reach a fairly high speed of descent, and the impact with the bottom will affect it. As the ship sinks, it slowly fills with water until the weight of the ship and cargo exceeds the buoyancy. As more water enters the hull, the speed gradually increases until it reaches a maximum that depends on the hull's shape and net weight. Compression of the wood may further decrease buoyancy and increase sinking speed. If the depth is sufficiently great, most ships land with the keel down, because of a relatively low center of gravity and the hydrodynamic forces on the underwater hull. On the seabed, ships often tilt to one side.

Parts of the hull can be damaged because the pressure pours water in and air out of the hull, and parts of the hull that do not fill with water immediately may collapse. R. D. Ballard estimated that the *Titanic* sank with a speed of 40–50 km/hour (11–14 m/s), based on full-scale experiments with other ships. It is likely that the Titanic reached its maximum speed after several minutes and used less than ten minutes to sink to 3,800 m (Ballard, 1987). It is, however, difficult to calculate the maximum sinking speed of a ship, for several factors influence the descent. Weight, surface area, and shape all affect the speed, and instabilities reduce the speed and make the sinking force change direction.

If a typical ancient vessel carrying four hundred amphoras or a similarly sized early medieval Scandinavian trading vessel (ca 15 m long, 4.5 m beam, 25–30 tons capacity) is used as an example, the theoretical maximum sinking speed is 3–4 m/s. The maximum speed that a typical 40 m eighteenth-century sailing ship would sink to the seabed is 6–8 m/s. The net weight of a ship made of wood need not be very great, and there are several examples of ships sinking very slowly. Bascom calculated the terminal falling velocity of a wooden sailing ship loaded with three thousand amphoras to be 1.5 m/s. He argued that this would not lead to substantial damage. In Sweden, several eighteenth-century ships, wrecked and abandoned off the island of Gotland, later drifted across the Baltic Sea and sank on the east coast of Sweden (Rønnby, 1995).

One can also calculate in theory the time any sinking ship takes to reach maximum speed, and even more

Sailboat run aground (unknown photographer)

important the depth necessary to reach that speed. In theory, maximum speed can be reached within thirty seconds, or a depth less than 100 m. In a real situation, with the relative instability of the sinking ship and with forces working in several directions, it would probably take minutes and require several hundred meters of depth.

The physical impact of a ship hitting the seabed at high speed can collapse parts of the hull. In addition, parts of the ship are forced down into the mud, creating a mud cushion that immediately buries parts of the ship. It is also possible that the water being forced behind the ship as it sinks creates a pressure wave that further destroys the hull. A ship sinking in shallow water is not influenced by the sinking speed or the impact with the seabed. Other factors are much more important, especially the physical contact between the ship and the shore, which continues after the ship has come to rest on the shallow seabed.

In deep water there is relatively little physical degradation after a ship reaches the bottom. Wave action, which creates violent to and fro motions near the bottom in shallow water, decreases to zero in depths greater than 100 m. This means that there is no strain on the hull planking and no sand abrasion caused by moving water. Deepwater currents are also too slow to shift the wreckage or scatter objects. Because the seabed is stable, out of reach of waves and with the little current, a wreck cannot trap waterborne sediments, so the parts that are buried remain stable, whereas the exposed parts have small chances of survival. In addition, the effect of water pressure is usually insignificant, for most materials are only slightly compressible; only if there is an enclosed air space does pressure cause major damage. The wood becomes waterlogged, with the tiny air spaces in the wood flooded, but the shape of the ship remains the same. On the other hand, pressure increases with depth by 1 atm per 10 m, so the enormous pressure at great depths may have a greater influence on deep shipwrecks.

In total, ships sinking in deep water are subject to different forces than those sinking in shallow water, and in general those acting on a ship in deep water are often less destructive. The forces that have most affected deepwater wrecks are those on the surface that caused the ship to sink, those acting during the sinking process, and the impact with the seabed.

Man-made Interference

Man-made interference with wreck sites is common in shallow water (salvage, recreational diving) but rare in deep water. Other man-made formation processes like trawling, dredging, and construction work on the seabed are also less frequent in deep water.

Preservation of Cultural Remains in Deep Water

In such an immense body of water as the oceans, there are unlimited geographic and local variations in the

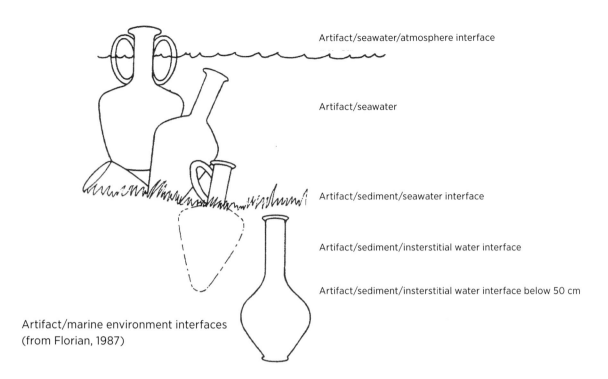

Artifact/marine environment interfaces (from Florian, 1987)

parameters of the marine environment. The part of the marine environment that is most significant with respect to artifact deterioration is that immediately associated with the artifact. This is usually a solid/liquid interface, such as artifact/seawater, or a solid/solid interface like artifact/sediment. Disintegration is caused by a complex combination of many physical, chemical, and biological processes. Predicting or interpreting changes that occur to artifact material involves an understanding of both the marine environment and the material (Pearson, 1987).

Only a few materials have a good chance of surviving without degradation under water, such as rock, some metals, ceramic materials, and glass. The survival of other materials depends on the condition of burial at the seabed. Generally, massive pieces have a better chance than small objects, and organic material, like wood, has usually completely disappeared. Given special conditions, though, nearly anything may survive for thousands of years in the sea.

Depth Zones

To distinguish possible differences in preservation conditions, we start by identifying the three principal depth zones: surface, pycnocline, and deep zone (Thurman, 2003). The surface zone is in contact with the atmosphere. It changes seasonally because of variations in precipitation, evaporation, cooling, and heating and has the warmest and least dense water in the ocean. The surface zone is 100–500 m deep, and its waters are usually well mixed by winds, waves, and temperature changes at the surface.

The pycnocline zone is around 1 km deep. The water density in the pycnocline zone changes markedly with depth. The top layer typically has a temperature of approximately 10°C and the lowest layer around 4°. Changes in either temperature or salinity can cause a marked change in density of the pycnocline. When changes are mainly caused by temperature, it is called a thermocline. Where salinity dominates, it is known as a halocline. A thermocline is important in the open ocean, where salinity changes little. Haloclines are more important in coastal ocean areas, where salinity changes dominate.

Below the pycnocline is the deep zone. Temperature and salinity in the deep ocean are unaffected by surface processes and act as conservation properties. Below the pycnocline, except in the high latitudes, there is little or no vertical movement, and currents are slow and poorly known. Deepwater currents differ markedly from currents at the surface. When a water mass reaches its appropriate density level, it spreads out, forming a thin layer. It takes less energy to move a parcel of water along a surface of constant density than across it, and currents therefore move horizontally along the surfaces of constant density. The densest water masses form only in the polar areas,

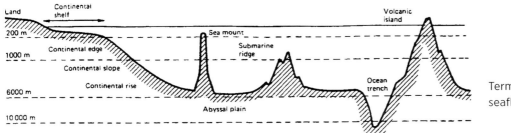

Terms applied to parts of the seafloor (from Morgan, 1990)

with the cooling of already highly saline waters. This cold, dense polar water sinks into the deeper part of all the major ocean basins and spreads along the ocean floor toward the tropics.

The principal boundary between continents and the deep ocean is the continental margin. Continental margins consist of continental shelves, continental slopes, and continental rises (Morgan, 1990). The continental shelves are the submerged upper parts of continental margins. From the shoreline the shelves slope toward the shelf break (average depth 130 m). There they join the steeper continental slope, which extends down to depths of 2–3 km. Off Antarctica the continental shelf breaks are about 500 m deep because the weight of the ice on the land has depressed the continent and the continental margin. The shelf is very deep around the Arctic Ocean for the same reasons. Continental rises occur at the base of the continental slopes. They slope gently seaward and connect with the deep ocean floor, usually named the ocean basins.

The deep ocean floor (deeper than 4,000 m) occupies about 30 percent of the earth's surface and consists of immense areas of flat seabed, called abyssal plains. In addition, common features of the deep sea are ridges, fracture zones, abyssal hills, and trenches. Trenches are the deepest parts of the ocean floor, typically 3–4 km deeper than the surrounding ocean floor. The greatest depth anywhere in the oceans is around 11,000 m, in the Mariana Trench, southeast of Japan. The average depth of the ocean is 3,800 m, but more than 80 percent is deeper than 3,000 m. Only a small percentage is deeper than 6,000 m.

There are obvious differences in the character of these various oceanic zones. The deep ocean differs considerably from shallow, coastal regions. The shallow, coastal areas are much more complicated than the open ocean. The many bodies of water—estuaries, fjords, lagoons—that are connected to the coastal ocean but partially isolated each have their own characteristics. Coastal currents, driven by the wind and large river discharges, generally parallel the shore and are much stronger than deepwater currents. Temperature and salinity changes can also occur in partially isolated coastal waters. The deep ocean, in contrast, is characterized by uniformity of conditions over great distances, although variations clearly do exist.

Deepwater archaeology, defined here as covering depths of 50–11,000 m, encompasses all depth zones and their various conditions. Two archaeological sites in different geographic locations in the ocean can have the same environmental influences. On the other hand, two sites in nearly the same geographic location can have different environmental parameters.

Chemical Processes

Several important chemical processes take place on an archaeological site. The main cause of chemical damage is oxidation. The ocean is not well oxygenated throughout its depths, since the oxygen is added to the upper layers of the seawater by absorption of air, and in regions of photosynthetic activity it is limited by light penetration. In the coastal ocean, the mixing process is rapid because of the wind, waves, and high-velocity coastal currents. Below the surface zone, dissolved oxygen can come only from subsurface waters. Dissolved oxygen is consumed at all depths of the ocean. It is lost mainly by exchange with the atmosphere and by utilization in aerobic respiration of plants and animals and decomposition of organic material by aerobic bacteria. The concentrations of dissolved oxygen also reflect the age of the water. There is less oxygen in old water. Thus, the deep waters of the North Pacific and Indian Ocean are most vulnerable to oxygen stress (Thurman, 2003).

Characteristic oxygen profiles of the main open oceans show that the surface maximum is the result of absorption at the atmosphere interface. This high concentration

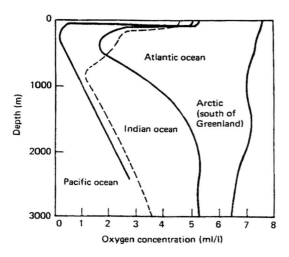

Characteristic oxygen profiles of the oceans (from Florian, 1987)

is then reduced by aerobic respiration in the upper region. The increase farther down is caused by slow aerobic respiration. At the sediment interface on the seabed, the dissolved oxygen may be depleted by aerobic biological decomposition of degradable organic material, but most deeper areas in the oceans are rich in dissolved oxygen.

Deep-sea sediments contain less organic carbon than shallow sediments. This carbon is the only reducing agent (a consumer of oxygen) supplied to sediments in any quantity. When there is a lot of organic carbon in the marine system, this can lead to deep-sea anoxia, an environment free of dissolved oxygen. In general oxygen consumption in the deep ocean is slow because of low temperatures and high pressure, which reduce the metabolic rates of organisms. But if the bottom contains a great excess of organic material that immediately uses up new oxygen, the result is a reducing environment with only limited chemical disintegration and the best conditions for preservation (Florian, 1987).

Even a slight decrease in salinity reduces the corrosion of metals. Minerals washed off land dissolve in rivers and travel to the sea suspended. When the seawater evaporates, these substances remain behind. Sodium chloride makes up more than three-quarters of the dissolved salts in seawater, followed by magnesium chloride, sodium sulfate, and several others. Salinity stays more or less constant between 32 and 37 parts per thousand in the open ocean, but it can be as low as 2 parts per thousand in areas of the cool, river-fed Baltic Sea and as high as 40 parts per thousand in the hot, dry-shored Red Sea. Salinity can also vary with the different depth zones.

Low temperatures help preservation by reducing the rate of chemical reaction. Temperatures vary according to the depth of water and its geographic location. Surface temperatures range from 29°C in the Red Sea in the summer to ⁻2°C in the polar seas in winter, the temperature at which seawater freezes. Since oceans are heated from above, the temperature in the oceans drops sharply at the thermocline, and the average temperature is only 3–8°C. Preservation conditions are therefore favorable in deep water.

A protective layer of sediments also limits chemical processes. The impact of a sinking ship landing on a soft mud bottom throws much mud into suspension that slowly settles back to form a protective covering. This mud layer may lead to an anaerobic/anoxic environment. It is therefore common to find that the upper exposed sections of a shipwreck have been attacked by chemical and biological processes and disappeared completely, while the lowest parts, including the keel, bottom planking, and lower ribs, have been covered with mud and survived. Heavy objects may also survive, since they sink into the soft bottom sediments and end up at a lower level.

Particle size, the degree of compaction, and the water content of the sediment determine the sediment's physical characteristics and the chemical and biological reactions within it. When sedimentation occurs, aerobic bacteria remove all the free oxygen, creating an anoxic environment (Florian, 1987). Several archaeologists (e.g., Muckelroy, 1978) have pointed to burial as the main determining factor in the survival of archaeological remains in shallow water, and this conclusion is also valid in deep water.

Sediment deposits are accumulations of minerals and rock fragments from land mixed with insoluble shells and bones of marine organisms and some particles formed through chemical processes in seawater. Sediment particles are classified according to grain size. Size is important because it determines where particles are transported given a certain current speed, and thus where they accumulate in the ocean. Large grains settle out near where they enter the ocean. Most riverborne particles are trapped by estuaries. Smaller particles can in theory be transported throughout the ocean. Typical sediment accumulation rates are found in table 7.1, where it can be seen that deepwater sites are less likely to be covered by sediments and thus less likely to survive than shallow sites, keeping other factors constant. In some extreme

Table 7.1 Typical sediment accumulation rates

Area	Average Accumulation Rate (cm per thousand years)
DEEP OCEAN	
Deep sea muds	0.1
Coccolith muds	1
MARGINAL OCEAN BASINS	10–100
CONTINENTAL MARGINS	
Continental shelf	30
Continental slope	20
Estuary, fjord	400
Delta	>1,000
WETLANDS	150

cases, turbidity currents consisting of dilute mixtures of sediments and water with a density greater than the surrounding water may lead to much higher sedimentation rates (Morgan, 1990). Turbidity currents are set off by slides, large discharges of riverborne sediments, or earthquakes and can reach velocities of at least 30 km/hour, resulting in erosion and quick sediment buildups that bury ocean bottom topography and possible cultural remains.

In summary, deepwater archaeological sites have the best chance of survival in areas with no oxygen, low temperatures, and low salinity. The deep ocean offers favorable conditions with respect to oxygen content and temperature. However, the most favorable conditions typically occur when artifacts have been covered by sediments. Because sedimentation rates in the deep sea are low, items that were not immediately covered by sediments after impact deteriorate and have limited chances of survival.

Physical Processes

The extreme cases of low preservation are wreck sites in shallow water where the currents and wave action are so strong that artifacts are physically tumbled and moved back and forth across the seabed. In such cases, mechanical damage and abrasion ensure that artifacts in general do not survive long unless they come to rest in crevices or depressions in the seabed. Growth of marine organisms on items can also be important, for they may act as a protective layer reducing metal corrosion. Such growth may also cement objects and the seabed together, making them less susceptible to physical damage. In shallow water, most wreck sites have been severely broken down.

Water movement across an artifact affects corrosion rates through metal erosion, destruction of protective films, or indirectly by changing the rate of oxygen supply to the cathodic reaction. Erosion of wood is also common, often accelerated by fungi, bacteria, and borers. But currents and wave action are predominantly shallow-water problems. The deep ocean is virtually unaffected by currents and wave action. Currents in the deep ocean travel quite slowly and are generally ineffective in producing rapid changes in conditions. Current speeds of 1–2 cm/s are typical. Surface currents, on the other hand, are mainly wind driven and dwindle at 100 m below the surface. Surface currents move at about 2 percent of the speed of the wind that causes them, so they can reach fairly high speeds.

In deep water, where the water depth is greater than half the wavelength (L) of a surface wave, the water is moved little by the wave's passage. The wavelength is limited to seven times the wave height, so the effect is quickly reduced with depth. Where the water depths are less than L/20, the bottom is strongly affected by waves. In addition to damaging a site by breaking up and moving objects, waves can also shift sediments and sand, covering and uncovering objects. Obviously wave-induced water and sediment motion is principally a feature of shallow water, and the scour and catchment of sediments that typically influence shallow sites are not common in the deep ocean.

A third disturbing force results from the gravitational attraction of the sun and the moon on water, resulting in tides. Tides are best expressed as waves with very long periods. Tidal currents are horizontal water movements associated with the rise and fall of the ocean surface caused by tidal forces. But the relationships between tides and tidal currents are not always obvious. Some coastal regions have no tidal currents, and a few have tidal currents but no tides. In restricted coastal seas the strength of a tidal current often depends on the volume of water that must flow through a small opening. In the open oceans, tidal currents exhibit patterns quite different from those in coastal areas and can constantly change direction. In general, tidal currents are strongest in the coastal regions and much weaker in the open ocean. In the deep sea, tidal currents typically move at about 5 cm/s, whereas in the shallow areas they may move at several meters per second. Currents in the deep sea still have a clear tidal component, but their effects on preservation conditions are limited, as can be understood from the above discussion. Thus these currents cause very little physical damage to sites below 100 m.

Biological Processes

Perhaps the most important bearers of disintegration of an archaeological site are biological processes. In the marine environment, biological organisms are restricted in their distribution by environmental parameters such as water salinity, temperature, depth, pressure, currents, nutrients, and sediment properties. The significant organisms are fouling assemblages, wood and stone borers, fungi, and bacteria (Florian, 1987).

A fouling assemblage is a community of animals and plants living on man-made objects in the sea. Nearly two thousand plant and animal taxa have been identified from fouling assemblages. All fouling organisms live in the upper levels of the sea at the depth of light penetration (photosynthetic zone, 200 m) and are dependent on oxygen for growth; they do not grow in anoxic environments or in sediments. Fouling may prevent or accelerate corrosion.

The wood-boring mollusks are the shipworms (*Teredinidae*) and the piddocks (*Pholodaceae*). Wood borers are clams that bore into wood by the rasping action of their clamshell grinders. When the borer larvae touch wood, they quickly bore in. Once inside, the organism follows the grain of the wood and leaves behind a tube about 0.8 cm in diameter. Research data from the U.S. Navy show that an entire wooden wreck could be devoured within 25–50 years (Bascom, 1991). Shipworms require at least 12 parts per thousand salinity to develop and are restricted to above 200 m. Piddocks have been found in depths down to 2,000 m. The main wood-boring crustacean is the gribble, which is limited to above 500 m (Florian, 1987). Only wood above the mudline is vulnerable to marine borer attacks. Not much is known about how wood borers develop. Fresh water from rivers reduces the salinity and decreases wood borer attacks. Certain types of wood, such as teak, suffer only limited attacks (Arbin, 1996). Wood borers have been active on every deepwater shipwreck that has been located in the open ocean.

The moment artifact material comes in contact with seawater, some degree of solubilization (waterlogging) occurs, with possible swelling and contamination from other materials such as iron salts. Water fills all the pore spaces in the wood, artifacts are immediately covered with bacteria, and the cycle of biodeterioration starts. Bacteria and other organisms take much longer to deteriorate organic material than do borers. Bacteria typically utilize artifacts as a nutrient, usually through aerobic oxidation of organic material. Marine fungi deteriorate organic materials such as wood, textiles, and cellulose. Fungi cannot grow in anoxic conditions and require dissolved oxygen levels of at least 0.30 ml/1 (Florian, 1987).

The deterioration rate is dependent on depth. It has been shown that bacteria are less efficient with depth. When the lost submersible *Alvin* was recovered from 1,500 m after eleven months on the seabed, the food aboard was remarkably well preserved (Stowe, 1996). It is likely that high pressure in combination with low water temperature causes lower growth rates and reproduction of bacteria. Anoxic conditions are created by local aerobic bacterial activity or abnormally rapid accumulation of organic matter under conditions of restricted circulation and high productivity. Aerobic biological activity is excluded in such conditions, but anaerobic bacteria still oxidize the organic material, even though the preservation of organic material appears enhanced in anoxic areas. Sulfate-reducing anaerobic bacteria can also lead to corrosion. During the reduction process, a slow transport of metal particles from metal surfaces causes rust. Beneath this rust, which may easily be removed, the surface is usually fine, but the metal thickness has been reduced, and this process will continue until the metal object is totally destroyed. This is a major problem on the *Titanic*. Some anoxic areas in the Pacific lack anaerobic bacteria.

Structures may survive if pinned down by the vessel's cargo and buried in sediments, avoiding attacks from aerobic bacteria and wood borers. In surface sediments that are rich in dissolved oxygen, aerobic bacteria are the dominant type of bacteria. When oxygen is depleted by aerobic oxidation of organic material, so-called facultative bacteria become dominant, whereas under anoxic conditions anaerobic bacteria are dominant. The population is greatest in the first 2.5 cm of sediment. Aerobic bacteria decrease at the 10 cm level; anaerobic bacteria extend to 62 cm. The depth of bacteria-free sediment is commonly taken to be 50 cm (Florian, 1987). Deterioration is also greater in the upper layers of sediment than in the overlying water, and metals, glass, and wood all show acceler-

Borer attacks on wood
(David Tuddenham, NTNU Vitenskapsmuseet)

ated deterioration at the mudline compared to material in the overlying water or deeply buried.

The deepwater environment receives biogenic debris derived from planktonic organisms. This material links the top and bottom of the ocean and provides an explanation for many apparent seasonal changes in the appearance of the deep seafloor and for the seasonal reproduction of some species of deep-sea benthic animals. All major groups of animals in the shallow-water benthos are present on the deep bottom, but there they are generally smaller than their shallow-water relatives. It has also been found that animals living in the deep sea are one hundred to one thousand times less abundant than in shallow water (Tyler et al., 1996).

Animals are the cause of another site formation process. In a muddy substrate there is ample evidence of burrowing openings, and deep-sea animals can modify the sediment in a bewildering variety of ways. Bottom-dwelling organisms in particular are responsible for burial of material, creating new stratigraphic relationships within a deposit and threatening the stability of material once buried. In some cases the annual turnover of sediment caused by burrows can be in the range of 6–8 kg/m^2, and tunnels can be 20–30 cm deep. The extent of such activity is primarily influenced by the suitability of the substrate. Burrows are caused by several organisms, among them worms, crustaceans, and fish. There are clearly different effects from large burrows compared to the higher densities of small, individual excavations (Ferrari and Adams, 1990). Anoxic conditions in sediments are thought to effect the best preservation, but burrows may create aerobic conditions by reexposing wood to marine borers. Burrowing is also known to change the stratigraphic position of objects. Burrowing activity ceases with sediment depth, since these organisms cannot live deeper than 50–60 cm in the sediments. Burrowing may also lead to mechanical weakening through loss of support and scratching of surfaces.

Large sections of structure coming to rest on the seabed gradually affect the environment. Wreck sites are known to increase the number and diversity of habitats in an area, and a cargo of organic material clearly creates a large food potential that increases feeding-related burrowing activity in the area. An individual artifact can be expected to be less influenced by biological processes than large sections of the ship, which are almost immediately affected. Deterioration caused by fouling assemblages, wood and stone borers, fungi, and bacteria, as well as by burrowing caused by animals, modify and destroy sites. In general, most biological processes are present in the deep sea but have less influence than in shallow water. In some cases anoxia, which eliminates aerobic bacteria and animals, occurs in deep water and leads to the most favorable conditions, where most biological and chemical influence is avoided. Similar conditions are found in both deep and shallow water when the remains are buried under 50 cm of sediments.

High-potential Areas

It is clear from the previous discussion that some areas have better preservation conditions than others. In general, the deep ocean has more preservative conditions than shallow areas, although there are clearly exceptions. Anoxic conditions represent the most preservative factor. The deep waters of the North Pacific (anoxic conditions and no anaerobic bacteria) and the Indian Ocean (anoxic conditions) are areas of high potential for excellent preservation conditions. Some cold areas also have favorable conditions, because of the very low temperatures and low salinity, which seem to be more influential than oxygen content, which is typically high in cold areas.

During the latest Quaternary glaciation, the Black Sea became a giant freshwater lake. When the Mediterranean rose to the level of the Bosphorus sill, about 7150 years BP, saltwater poured through this spillway to refill the lake and submerge, catastrophically, more than 100,000 km^2 of exposed continental shelf (Ballard, Coleman et al., 2000). At the breakthrough point the floodwaters would have been tens of meters deep very quickly and moving at an extremely high velocity, as great as 100 km/hour at the narrowest point of the flow. Sea level would have risen about 15 cm/day every day for two hundred to three hundred days. This event has been proposed as a possible historical source for the biblical account of Noah's Flood as well as other flood myths from ancient societies in the Near East. The pre-Flood Black Sea would have been an oasis for Neolithic peoples living on its shores. The permanent drowning of this vast terrestrial landscape may have accelerated the dispersal of early Neolithic foragers and farmers into the interior of Europe, Anatolia, the Aegean, and the Levant at that time.

Although the archaeology, mythology, and biblical studies of Noah's Flood are speculation at this time, they draw on new scientific evidence to ground their speculations. Joint Russian-U.S. projects have revealed paleochannels and erosion surfaces at depths of 160 m with a uniform overburden draping the subsurface topography, indicating a rapid sea transgression. Additional evidence comes from seafloor cores, which have recovered mollusk shells from three different species at 49–123 m. Radiocarbon dates of the shells were around 7500 years BP.

Presently, the Black Sea surface salinity is approximately 18 parts per thousand, increasing to 22 parts per thousand at depth. Saline water from the Mediterranean Sea enters the Black Sea along the bottom of the Bosphorus and spreads across the Black Sea, but it is limited to the upper 500 m and does not reach the bottom. The less saline seawater flows from the Black Sea into the Mediterranean above this counterflowing saline layer. The water column is characterized by the absence of oxygen and elevated concentrations of hydrogen sulfide and methane from about 100 m to the bottom at depths below 2,000 m.

The anoxic conditions are created because oxygen is consumed by oxidation of sinking organic matter, and a shallow, sharp salinity-determined density gradient prevents exchange of oxygen between the surface and deep water.

The anoxic conditions of the deep Black Sea may perfectly preserve ancient and historical evidence of pre-Flood habitation and post-Flood seafaring, offering archaeologists an unparalleled opportunity to study the past in a relatively undisturbed state. With anoxic conditions deeper than 100 m, the Black Sea is ideally suited for deepwater archaeological investigation.

The Baltic Sea has also been an important area of Europe's most dominating trading and military powers since prehistoric times. The Baltic is only one-sixth the size of the Mediterranean, with a mean depth of only 65 m and only a narrow channel between Denmark and Sweden linking it with the Atlantic. This threshold controls the flow of water in and out of the Baltic Sea. Low-density water that has been diluted by freshwater inflow from rivers pours out through this channel, flowing over the high-density, saltier water in the North Sea, which cannot readily push in underneath. Thus the Baltic is extremely brackish, and the low salt content helps keep out wood-boring shipworms. This has created some remarkable ship finds, like the *Vasa* (Franzen, 1966), and on several recently discovered shipwrecks in the Baltic the hulls are more or less intact. Similar conditions can be found in lakes and rivers and in the Great Lakes in the United States and Canada.

Some fjords also offer excellent preservation conditions. The chief characteristic of a fjord is the presence of a shallow sill at the seaward end. Freshwater outflow is limited to the depths above the sill, while incoming saltwater flows over the sill and fills up the basin behind the sill. Saltwater of oceanic origin frequently becomes stagnant because its circulation is limited by the shallow sill depth. Thus, in many cases the deep water in a fjord is very deficient in oxygen, making conditions for preservation favorable.

Again, though, we should not associate archaeological value only with preservation. Important discoveries may just as well come from poorly preserved but more unusual sites along the coast or in the open ocean. If asked, archaeologists also mention shipping routes and sites of great naval battles as areas of high potential archaeological value.

EIGHT

Deepwater Archaeology: Law and Ethics

Most European countries insist that underwater cultural heritage belongs to the national government, with few rewards to the finder. In 1987 the U.S. Congress passed the Abandoned Shipwreck Act, which holds that all shipwrecks under water are the property of the U.S. state in whose waters they are discovered. Some argue that such approaches discourage responsible private parties from even looking, while the coastlines are still being clandestinely pillaged by individuals who do not report their discoveries since they know they will lose what they find.

UNESCO Convention on the Protection of Underwater Cultural Heritage

At present there is no international legal instrument that adequately protects underwater cultural heritage, which is increasingly threatened by construction work, advanced technology that enhances the identification of and access to wrecks, exploitation of marine resources, and commercialization of efforts to recover underwater cultural heritage. This has led to the irretrievable loss of a vast part of our collective cultural heritage.

In response, UNESCO member states prepared the UNESCO Convention on the Protection of Underwater Cultural Heritage, which provides basic protection beyond the territorial seas of coastal states and aims to avoid and resolve jurisdictional issues involving underwater cultural heritage. The convention considers underwater cultural heritage an integral part of the cultural heritage of humanity and an important element in the history of peoples, nations, and their relations with each other. It states that exploration, excavation, and protection of underwater cultural heritage must be based on an application of special scientific methods, suitable techniques and equipment, as well as high degrees of professional specialization. The aim of the convention is to impose these standards for underwater scientific work and to make sure that the work is approved by responsible authorities. Some 350 experts from more than ninety countries worked for four years to finalize the draft document, which covers all traces of human existence having a cultural, historical, or archaeological character that have been partially or totally under water, periodically or continuously, for at least one hundred years.

The UNESCO Convention on the Protection of the Underwater Cultural Heritage was adopted in 2001 by the plenary session of the 31st General Conference by eighty-seven affirmative votes. Four states voted against (Norway, Russian Federation, Turkey, Venezuela) and fifteen abstained from voting (Brazil, Colombia, Czech Republic, France, Germany, Greece, Guinea-Bissau, Iceland, Israel, Netherlands, Paraguay, Sweden, Switzerland, United Kingdom, Uruguay). The United States could not vote because it is not a member of UNESCO, but as an observer it made a strong statement against the convention. The next step for member states is the deposit of an instrument of ratification, acceptance, approval, or accession. The convention can be enforced only after twenty countries have become party to it.

Since the convention would apply only to signatories, it is likely that additional efforts are needed to protect heritage in international waters, since the countries that possess the vast majority of technological access to the world's oceans will not be part of this effort.

Legal Standing of Underwater Archaeology in the United Kingdom

In 1992 the responsibility for marine archaeology was adopted by the Department of National Heritage, which is responsible for the Protection of Wrecks Act 1973 (Momber, 2000). Artifact retrieval in the United Kingdom has a long history and was being practiced from at least the seventeenth century. Ship salvage was seen as an achievement, not a threat, and this is reflected in the legislation. The main concern for shipowners was the retrieval of their property after wrecking. This concern is reflected in the Merchant Shipping Act 1894, revised in 1995. The act recognizes the rights of original owners and stipulates that any vessels or their content retrieved from tidal waters should be reported to the Receiver of Wreck. Efforts are to be made to locate the original owner, who has one year to claim the material, following which the Crown gains entitlement to the unclaimed wrecks. Unfortunately, the law indirectly encourages retrieval of goods from sites in lieu of a salvage award.

The Protection of Wrecks Act is used to protect wreck sites of historical, archaeological, or artistic importance from unauthorized interference. To date, however, only forty-nine wrecks around U.K. waters are protected, a very small number compared to the 32,000 records of wreck incidents compiled by the Maritime Section of the National Maritime Record in the Royal Commission on the Historical Monuments of England.

Other legislation that protects wrecks is the Protection of Military Remains Act 1986. The act was passed to protect the remains of artifacts or vessels used in military service that have been crashed, sunk, or stranded. Interestingly, the Royal Navy initially issued salvage rights to the HMS *Sussex* to Odyssey Marine Exploration, following the long British tradition of goods retrieval rather than protection of cultural heritage, but because of public protest and interference by other countries Odyssey has been forced to postpone further work on the project until legal and diplomatic issues have been resolved.

More recent developments include national and EU efforts to not grant planning permissions until an adequate environmental impact assessment has been conducted ahead of seabed development. This is a common procedure in European countries to protect underwater heritage, although it focuses on industry development and avoids dealing with treasure hunting.

Legal Standing of Underwater Archaeology in the United States

The National Historic Preservation Act of 1966 was enacted to recognize that the nation is "founded upon and reflected in its historic heritage." By passing this act into law, Congress proclaimed that "the preservation of this irreplaceable heritage is in the public interest so that its vital legacy of cultural, aesthetic, inspirational, economic, and energy benefits will be maintained and enriched for future generations." A series of federal laws enacted primarily in the 1960s provided for the creation of state protection of underwater heritage. The Abandoned Shipwreck Act is applied to state waters, the National Marine Sanctuary Act protects shipwrecks in national marine sanctuaries, and the Antiquities Act is limited to marine protected areas such as national seashores (Varmer and Blanco, 1999).

The continental shelf remains under the jurisdiction and control of the federal government. The Minerals Management Service (MMS), as a federal agency, is obligated under Section 106 of the National Historic Preservation Act to consider the effect of its actions on properties deemed important enough to be eligible for the National Register of Historic Places. As a result, the MMS has pursued a policy, first through lease stipulation and then through regulation, of requiring that lessees do not "unnecessarily jeopardize or harm a cultural resource which has been identified or is believed to exist." As the oil and gas industry moves its operations farther and farther from shore and into ever-deeper water, the MMS faces new challenges to protect significant archaeological resources on the seabed. The MMS estimates that there are well over four thousand historical shipwrecks in the Gulf of Mexico which, if discovered, could meet these criteria.

Naval vessels belonging to the United States, regardless of location, fall under the auspices of the Underwater Archaeology Branch of the Naval Historical Center through the Sunken Military Craft Act.

Norwegian Law for the Protection of Cultural Heritage

The Norwegian law for protection of cultural heritage is one of the strictest in the world. In principle, the Norwegian state is the legal owner of all cultural remains found under water. The Norwegian law for the protection of cultural heritage of 1963 included shipwreck sites older than one hundred years but did not consider the cargo or single objects. After the discovery of a treasure wreck in 1972, the law was amended to include these items. In the latest revision (1978), the law states that before any construction work can be initiated under water it is necessary to carry out marine archaeological investigations in areas that may be damaged. The developer is usually responsible for the costs of these investigations, but if the costs are extraordinarily high the Norwegian government can pay for the investigations. The result is that cultural remains on land and under water are protected in the same way.

Marine archaeological investigations are carried out by five institutions each responsible for a section of the Norwegian coastline. Every year many hundred cases are considered by these five institutions. In the past fifteen years a much stricter practice has been possible thanks to an increase in available personnel, experience, and increased awareness of marine archaeology in general. The result has been that oil companies operating offshore must also consider protection of cultural remains on the seabed to make sure that shipwrecks and ancient settlements are not damaged by their construction work in deep water.

In the late 1980s the Norwegian state oil company, Statoil, and other companies started to explore the oil and gas reservoirs on Haltenbanken in the Norwegian Sea. As a result of finding the Heidrun oil and gas field, Statoil and Conoco planned to build a new gas pipeline from the offshore continental shelf to the shore. After several seasons of sonar and topographical seabed surveys, a potential pipeline route, expected to be the route of fewest physical obstacles and lowest cost, was selected. This 250 km pipeline route crosses over a rugged seabed with depths more than 300 m. Because the pipeline route passed close to the Haltenbanken fishing bank, the project was named the Haltenpipe Development Project (HDP).

During the initial surveys through the coastal approaches, at Ramsøy fjord between the islands of Smøla and Hitra, it was discovered that the remnants of an old shipwreck might be situated close to the planned track. Further investigations showed that ship's cargo had been found in this area earlier, near the shore (see chapter 3).

When this incident was brought to Statoil's attention in 1991 during the initial inspection of the pipeline track, not much was known about the ship—its identity, the purpose of its journey, its cargo. The oil company was interested in completing the pipeline without external interference, but NTNU, who is responsible for the protection of cultural remains in this part of Norway, claimed that remnants of the wreck could be damaged by the pipeline and demanded that special investigations be carried out before the oil company could lay its pipeline.

In all types of engineering construction projects, Statoil, as all other companies, naturally keeps the costs as low as possible. In the Haltenpipe project, only costs necessary for the safe and legally sound completion of the project were to be covered.

Prior to the HDP, Statoil had already completed several offshore field development projects, including the construction of four major offshore pipelines, totaling more than 5,000 km of pipe, in depths down to 540 m. Through this significant experience, the company knew that detailed underwater mapping, inspection, and surveying were crucial to success.

During the initial laying of Statoil's first offshore pipeline, Statpipe, between the Statfjord field and western Norway, the prelay route inspection in front of the laybarge uncovered intact World War I mines. Even though these objects had been detected with sidescan sonar imagery during the initial survey, no intact mines had been expected, so the objects had been interpreted as boulders. Naturally, the cost of holding up the laying operation for several days, for mine clearance, far superseded all survey expenditure for initial pipeline route surveying and inspection. Similarly, unplanned marine archaeological investigations of the pipeline track late in the project could both delay the pipelaying operation and increase project costs.

To this point, compared to the normal case on land, there had been little legal or government involvement in most offshore underwater development projects. Besides fisheries and military interests, there were only rarely other rights-of-way to consider. But in the HDP project, NTNU claimed that Statoil was also responsible for protecting any old shipwreck discovered during the initial pipeline route survey and inspection. Because Statoil had not previously experienced similar considerations on the pipeline projects farther south in the North Sea, it at first attempted to claim no responsibility for marine archaeology.

By tradition, onshore archaeology had been treated differently by Statoil. Whereas the claims for marine archaeological investigations came as a surprise to Statoil, it treated onshore archaeology as a routine consideration, which according to Norwegian law had to be taken seriously. During the Europipe onshore route planning over German ground, such investigations also had to be performed there, since Germany enforced its laws dealing with onshore archaeological sites.

NTNU claimed that the law also applied offshore if a project threatened sites considered by marine archaeologists to be of special heritage interest. Since the pipeline could damage cultural remains, NTNU claimed that Statoil had an obvious responsibility to make sure that proper investigations were undertaken prior to detailed construction planning.

From Statoil's point of view, the government, including the ministries of defense, nature conservation, science and education, and the Norwegian Petroleum Directorate, had already approved the HDP routing and given their consent to construction. Against this view, NTNU argued that, even though Statoil had been given the right to build the pipeline, this could not be done without a consideration of the possible effects on cultural remains on the seabed—which meant that Statoil had to accept the added costs of an investigation to establish whether the pipeline could destroy cultural remains.

NTNU claimed that it had a mandate to halt any construction in locations of archaeological importance through Norwegian law, but that claim was at first not taken seriously by Statoil. Statoil then considered two options: ignore the NTNU claim and await lawsuit, perhaps risking a government halt to laying operations, a situation Statoil thought unlikely; or cooperate with NTNU. Statoil decided to cooperate and to include the extra costs into the project.

From previous experience with academic research institutions in Norway, Statoil had managed to develop good relationships in which both Statoil and the institutions often gained in knowledge, experience, and public esteem, thus creating a win-win situation. The arguments needed internally in Statoil to defend such (academic) expenditure, often considered by seasoned construction engineers a waste of money, were rather delicate. However, in Statoil internal documents and statements on etiquette, environment, nature, and public relations, one can find ample scope for assistance in measures to protect cultural heritage, just as there is stimulation from top management to protect nature and environment in any possible way. Therefore, the agreement with NTNU was not made in fear of the authorities holding up the laying operations but rather in the spirit of sustainable development, research, and basic respect of our heritage.

In total, the HDP costs for the marine archaeological work totaled about 0.3 percent of the total construction costs. Compared to a Norwegian law that stipulates that 5 percent of a building's total cost shall be spent on embellishment, such an expenditure on marine archaeology and related items is small. Philosophically and ideally, Statoil and other oil companies sooner or later have to include dimensions of heritage and sustainable development into all projects. There is, naturally, in all engineering projects a fear of spending money on nonserious activities and perhaps on profit seekers. Marine archaeology is, however, considered important and justified.

Because underwater construction work is still in its infancy, extensive rules and regulations comparable to those on similar onshore work are still mostly lacking. But it may be just a matter of years before the seafloor becomes the domain of the general public and all our underwater installations and activities become visible to the media. Therefore, we should already be introducing sound, aesthetic standards, similar to those we impose on land, so that we can be proud of, rather than fear, future evaluations of underwater constructions. In Norway, the HDP project led the way for several additional deepwater archaeological projects, culminating with the Ormen Lange project, the world's most advanced deepwater archaeology project and the first deepwater excavation of a shipwreck site by an archaeological institution. This could not have been achieved without the support of the oil companies and their basic respect of underwater heritage. It proves that the field of deepwater archaeology will continue to develop with the aid of scientists, researchers, and responsible companies and individuals.

REFERENCES

Adams J. 1996. The Kravel project: Archaeology beyond the air range. Third Underwater Science Symposium. Society for Underwater Technology, Bristol, U.K.

Adams, J., and Rönnby, J. 1996. *Furstens fartyg.* Sjöhistoriska museet, Stockholm.

Akal, T., R. D. Ballard, and G. F. Bass (eds.). 2004. *The application of recent advances in underwater detection and survey techniques to underwater archaeology.* Uluburun Publishing.

Alberg, D., J. Johnston, E. Silver, and K. Trono. 2008. *Monitor National Marine Sanctuary Condition Report 2008.* Office of National Marine Sanctuaries, National Oceanic and Atmospheric Administration, U.S. Department of Commerce.

Allwood, R. L. 1990. Diving and underwater vehicles. In N. Morgan (ed.), *Marine technology reference book.* Butterworths.

Andersen, J. 1969. A new technique for archaeological field measuring. *Norwegian Archaeological Review,* no. 2: 68–77.

Arbin, S. von. 1996. Havets husbockar—marine träborrande organismer som formations process. *Marinarkeologisk tidskrift,* no. 1: 12–14.

Arnoux, G. 1996. Physiological effects of diving on divers. 3rd Underwater Science Symposium. Society for Underwater Technology, Bristol, U.K.

Atauz, A. D. 2008. *Eight thousand years of maltese maritime history: Trade, piracy, and naval warfare in the central Mediterranean.* University Press of Florida.

Ayuso, V. M. G., and B. R. Bernal. 1992. *Catálogo de las Ánforas Preromanas.* Dirección General de Bellas Artes y Archivos.

Ballard, R. D. 1987. *The Discovery of the Titanic.* Madison.

———. 1993. The Medea/Jason remotely operated vehicle system. *Deep Sea Research I* 40.8: 1673–87.

———. 1998. High-tech search for Roman shipwrecks. *National Geographic,* April: 32–41.

———. 2007. Archaeological oceanography. In *Remote sensing in archaeology: Interdisciplinary contributions to archaeology.* Springer.

——— (ed). 2008. *Archaeological oceanography.* Princeton University Press.

Ballard, R. D., D. F. Coleman, and G. D. Rosenberg. 2000. Further evidence of abrupt Holocene drowning of the Black Sea Shelf. *Marine Geology* 170.3–4: 253–61.

Ballard, R. D., F. T. Hiebert, D. F. Coleman, C. Ward, J. Smith, K. Willis, B. Foley, K. L. Croff, C. Major, and F. Torre. 2001. Deepwater archaeology of the Black Sea: The 2000 season at Sinop, Turkey. *American Journal of Archaeology* 105.4: 607–23.

Ballard, R. D., A. M. McCann, D. Yoerger, L. Whitcomb, D. Mindell, J. Oleson, H. Singh, B. Foley, J. Adams, and D. Picheota. 2000. The discovery of ancient history in the deep sea using advanced deep submergence technology. *Deep-Sea Research I* 47.9: 1591–1620.

Ballard, R. D., L. E. Stager, D. Mater, D. Yoerger, D. Mindell, L. L. Whitcomb, H. Singh, and D. Piechota. 2002. Iron Age shipwrecks in deep water off Ashkelon, Israel. *American Journal of Archaeology* 16.2: 151–68.

Barto, Arnold J. 1996. Magnetometer survey of La Salle's ship the Belle. *International Journal of Nautical Archaeology* 25.3/4: 243–49.

Barto Arnold J., III, T. J. Oertling, A. W. Hall. 1999. The Denbigh project: Initial observations on a Civil War blockade-runner and its wreck-site. *International Journal of Nautical Archaeology* 28.2: 126–44.

Bascom, W. 1976. *Deep water ancient ships.* David and Charles.

———. 1991. Deepwater salvage and archaeology. In P. Throckmorton (ed.), *The sea remembers: Shipwrecks and archaeology.* Smithmark.

Beasley, T. F. 1991. The City of Ainsworth: An ROV analysis of a 19th-century lake sternwheeler. Underwater Archaeology Proceedings from the Society for Historical Archaeology Conference, Richmond, Va.

Bell, C., M. Bayliss, and R. Warburton. 1994. *Handbook for ROV pilot/technicians.* Oilfield Publications.

Bell, D. L., and T. A. Nowak. 1993. Some quality control considerations of remote sensing survey for archaeological sites. Underwater Archaeology Proceedings from the Society for Historical Archaeology Conference, Kansas City, Mo.

Bellingham, J. G. 1992. New oceanographic uses of autonomous underwater vehicles. *MTS Journal* 31.3: 34–47.

Bjørkvik, E. 1974. Russerforliset i Kvenværet i 1760. Tradisjon og dokumenter. *Årbok for Fosen.* Fosen Historielag.

Blot, J.-Y. 1996. *Underwater archaeology: Exploring the world beneath the sea.* Thames and Hudson.

Boyce, J. I., E. G. Reinhardt, and A. Raban, 2004. Magnetic detection of buried harbour structures and mooring sites in King Herod's Harbour, Caesarea Maritima, Israel. In T. Akal, R. D. Ballard, and G. F. Bass (eds.), *The application of underwater detection and survey techniques in archaeology.* Institute for Nautical Archaeology and Uluburun Press.

Bowens, A. (ed.). 2008. *Underwater archaeology. The NAS guide to principles and practice.* 2d ed. Wiley-Blackwell.

Bryn, P., M. E. Jasinski, and F. Søreide. 2007. *Ormen Lange: Pipelines and shipwrecks.* Universitetsforlaget, Oslo.

Cain, E. 1991. Naval wrecks from the Great Lakes. In P. Throckmorton (ed.), *The sea remembers: Shipwrecks and archaeology.* Smithmark.

Caravale, A., and I. Toffoletti. 1977. *Anfore antiche.* Orvieto.

Caverly, R. D. 1988. The application of SHARPS and CAD technology on the Yorktown shipwreck archaeological project. Underwater Archaeology Proceedings from the Society for Historical Archaeology Conference, Reno.

Cederlund, C. O. 1988. An effort to classify and describe the Swedish marine archaeological remains. Underwater Archaeology Proceedings from the Society for Historical Archaeology Conference, Reno.

Chance, T. S., A. A. Kleiner, and J. G. Northcutt. 2000. The autonomous underwater vehicle (AUV): A cost-effective alternative to deep-towed technology. *Integrated Coastal Zone Management* 1: 65–71.

Christensen, A. E. 1989. *Norsk sjøfart—begynnelsen fram til middelalderens slutt.* Norsk Sjøfart, Dreyers forlag.

Clyens, S and J. W. Automarine. 1996. A new method of raising objects from the seabed. *Underwater Systems Design* 18.6: 7–11.

Coleman, D. F., and R. D. Ballard. 2004. Archaeological oceanography of the Black Sea. In T. Akal, R. D. Ballard, and G. F. Bass (eds.), *The application of recent advances in underwater detection and survey techniques to underwater archaeology.* Uluburun Publishing, Istanbul.

Coleman, D. F., R. D. Ballard, and T. Gregory. 2003. Marine archaeological exploration of the Black Sea. Oceans 2003 Conference Proceedings, San Diego, Calif.

Coleman, D. F., J. B. Newman, and R. D. Ballard. 2000. Design and implementation of advanced underwater imaging systems for deep sea marine archaeological surveys. Oceans 2000 Conference Proceedings, Providence, R.I.

Conte, G., S. Zanoli, D. Scaradozzi, L. Gambella, and A. Caiti. 2007. Data gathering in underwater archaeology by means of a remotely operated vehicle. XXI International CIPA Symposium, Athens, Greece.

Couet, H.-G. de, and A. Green. 1989. *The manual of underwater photography.* Verlag Christa Hemmen.

Crawford, A. C. 1995. The Francois Vieljeux: Salvage in 1250 meters. In *Man-made objects on the seafloor: discovery, investigation and recovery.* Society for Underwater Technology, London.

Croizeau. I. 1996. L'épave de Sud-Caveau I: deux mille ans sous la mer. *Eureka*, no. 11.

Crumlin-Pedersen, O. 1991. Ship types and sizes AD 800–1400. In *Aspects of maritime Scandinavia AD 200–1200.* Proceedings of the Nordic Seminar on Maritime Aspects of Archaeology, Roskilde 1989. Viking Ship Museum.

Cullen, V. (ed.). 1989. Jason Project and Jason Foundation for Education. Annual report, Woods Hole Oceanographic Institution, Woods Hole, Mass.

Dane, A. 1990. Deep quest. *Popular Mechanics,* January: 56–59.

Dean M., B. Ferrari, I. Oxley, M. Redknap, and K. Watson (eds.). 1995. *Archaeology underwater: The NAS guide to principles and practice.* Nautical Archaeology Society.

Debrule, P., E. Saade, and A. Palmer. 1995. Laser line scan. In *Man-made objects on the seafloor: Ddiscovery, investigation and recovery.* Society for Underwater Technology, London.

Delaporta, K., M. E. Jasinski, and F. Søreide. 2006. The Greek-Norwegian deepwater archaeological survey. *International Journal of Nautical Archaeology* 35.1: 79–87.

Dobbs, C. T. C. 1995. The raising of the Mary Rose: Archaeology and salvage combined. *Man-made objects on the seafloor: Discovery, investigation and recovery.* Society for Underwater Technology, London.

Dobson, N. C. 2005. Developmental deep-water archaeology: A preliminary report on the investigation and excavation of the 19th-century side-wheel steamer SS Republic, lost in a storm off Savannah in 1865. Oceans 2005 Conference Proceedings, Washington, D.C.

Drap, P., and L. Long. 2002. *Towards a digital excavation data management system: The "Grand Ribaud F" Estruscan deep-water wreck.* Association for Computing Machinery.

Duffy, H. M. 1992. The high-tech world of Seahawk. *Treasure* 23.2: 38–43.

Dybedal, J., A. Løvik, and O. Malmo. 1986. The parametric array source and application of signal processing. Ultrasonic Symposium, Denver.

Einarson, L. 1990. Kronan: Underwater archaeological investigations of a 17th-century man-of-war. The nature, aims and development of a maritime cultural project. *International Journal of Nautical Archaeology* 19.4: 279–97.

Ekberg, G. 1997. Sonarkörning I Stockholms inre vatten. *Marinarkeologisk Tidskrift*, no. 3.

El-Hakim, S. F., A. W. Burner, and R. R. Real. 1989. Video technology and real-time photogrammetry. In I. Karara (ed.), *Non-topographic photogrammetry.* American Society for Photogrammetry and Remote Sensing.

Ewins, N. J., and D. A. Pilgrim. 1997. The evaluation of PhotoModeler for use under water. 4th Underwater Science Symposium, Society for Underwater Technology, Newcastle upon Tyne.

Farb, R.M, 1992. Computer video image digitisation on the USS Monitor: a research tool for underwater archaeology. Underwater archaeology proceedings from the Society for Historical Archaeology conference, Kingston, Jamaica: 100–104.

Fastner, J., F. Gaustad, and J. Kloster. 1976. Fregattskipet Perlen. 16. mars 1781. Utgraving 1975. DKNVS. Museet and Trondhjems Sjøfartsmuseum. Rapport, Marinarkeologisk serie 1976:1.

Fastner, J., and K. Sognnes. 1983. Seismisk registreringsutstyr i marinarkeologien—utprøving av utstyr i Trondheimsfjorden 1981 og 1982. Rapport, arkeologisk serie 1983:5. Årshefte 1983: 47–69.

Ferrari, B., and J. Adams. 1990. Biogenic modification of marine sediments and their influence on archaeological material. *International Journal of Nautical Archaeology* 19.2: 139–51.

Fish, J. P., and H. A. Carr. 1990. *Sound underwater images: A guide to the generation and interpretation of side scan sonar data.* Lower Cape Publishing.

———. 2000. *Sound reflections: Advanced applications of side scan sonar.* Lower Cape Publishing.

Florian, M.-L. E. 1987. The underwater environment. In C. Pearson (ed.), *Conservation of marine archaeological objects.* Butterworths.

Flow, J. 1996a. Excavation of a deep water shipwreck site. Seahawk Deep Ocean Technology.

———. 1996b. Seahawk I. Dry Tortugas wrecksite, preliminary report. Seahawk Deep Ocean Technology.

Foley, B. P., and D. A. Mindell. 2002. Precision survey and archaeological methodology in deep water. *Enalia, the Journal of the Hellenic Institute of Marine Archaeology* 6: 49–56.

Ford, B., A. Borgens, W. Bryant, D. Marshall, P. Hitchcock, C. Arias, and D. Hamilton. 2008. Archaeological excavation of the Mardi Gras shipwreck (16GM01), Gulf of Mexico Continental Slope, OCS Report, MMS 2008-037.

Franzen, A. 1966. *The warship Wasa.* Norstedts.

Frey, D., F. D. Hentschel, and D. H. Keith. 1978. The Capistello wreck excavation, Lipari Aeolian Islands. *International Journal of Nautical Archaeology and Underwater Exploration* 7.4: 279–300.

Gallimore, D., and A. Madsen. 1996. *Remotely operated vehicles of the world.* Oilfield Publications.

Garrison, E. G. 1989. A diachronic study of some historical and natural factors linked to ship-wreck patterns in the northern Gulf of Mexico. Underwater Archaeology Proceedings from the Society for Historical Archaeology Conference, Baltimore.

———. 1992. Recent advances in close range photogrammetry for underwater historical archaeology. *Historical Archaeology* 26: 97–104.

Gawronski, J. 1992. Functional classifications of artifacts of VOC-ships: The archaeological and historical practice. Underwater Archaeology Proceedings from the Society for Historical Archaeology Conference, Kingston, Jamaica.

Gearhart, R. 2004. Marine remote sensing: The next generation. Society for Historical Archaeology, 37th Conference on Historical and Underwater Archaeology, St. Louis, Mo.

Gifford, J. A. 1993. Videography and geographical information systems for recording the excavation of a prehistoric site. *International Journal of Nautical Archaeology* 22.2: 167–72.

Goddio, F. 1994. The tale of the San Diego. *National Geographic*, July: 34–57.

Greenough, J., P. Dart, and P. Holt. 1996. Recent developments in acoustic techniques used in marine archaeology. 3rd Underwater Science Symposium, Society for Underwater Technology, Bristol, U.K.

Hadjidaki, E. 1996. Underwater excavations of a late fifth century merchant ship at Alonnesos, Greece: The 1991–1993 seasons. *BCH* 120: 561–93.

Hagberg, B., J. Dahm, and C. Douglas. 2008. Vrak i Östersjön. Prisma, Stockholm.

Hansen, R. K., and P. A. Andersen. 1996. A 3D underwater acoustic camera: Properties and applications. *Acoustical Imaging* 22: 607–11.

Hardy, K. 1991. Return to the Titanic: The third manned mission. *Sea Technology,* December: 10–19.

Herdendorf, C. E., and J. Conrad. 1991. The S.S. Central America expedition: Hurricane gold. *Mariners Weather Log,* Summer: 4–23.

Hill, R. W. 1994. A dynamic context recording and modelling system for archaeology. *International Journal of Nautical Archaeology* 23.2: 141–45.

Hocking, C. 1969. *Dictionary of disasters at sea during the age of steam.* Lloyds's Register of Shipping.

Holt, P., A. Hildred, and P. Estaugh. 2004. Acoustic technology in historic wreck recovery. *Hydro International* 8.2: 6–9.

Hovland, M., M. E. Jasinski, and F. Søreide. 1998. Underwater construction projects vs. marine archaeology: How solving conflict saved old shipwreck. *Norwegian Oil Review* 24.7: 98–102.

Hughes Clarke, J. E. 1998. Detecting small seabed targets using high-frequency multibeam sonar. *Sea Technology,* June: 87–90.

Hummel, J. K. 1995. Improving ROV pilot performance using a multimedia system. *Underwater Systems Design* 17.2: 14–17.

Hutchinson, G. 1994. *The wreck of the Titanic.* National Maritime Museum, London.

Ingham, A. E. 1984. *Hydrography for the surveyor and engineer.* BSP Professional Books.

Irion, J. 2001. Cultural resource management of shipwrecks on the Gulf of Mexico Outer Continental Slope. Minerals Management Service, U. S. Department of the Interior.

Jasinski, M. E. 1994. Enhjørningen—om et dramatisk forlis med maritimarkeologisk etterspill. *Spor,* no. 2: 36–40.

———. 1995. Kong Øysteins havn på Agdenes. Forskningsstatus og revurderte problemstillinger. *Viking* 58: 73–104.

Jasinski, M. E., and F. Søreide F. 1997. Sjøfarten gjennom tusen år—marine undersøkelser i Trondheim havn. *Spor,* no. 1: 42–45.

———. 2001. Applications of remote sensing in Norwegian marine archaeology and management of underwater heritage. ICOMOS '99—World Congress on Monumental Heritage Conservation, Mexico City, Memorias del Congreso Cientifico de Arqueologia Subacuatica ICOMOS, Serie Arqueologia, Instituto Nacional de Antropologia e Historia, Cordoba.

———. 2003. The Norse settlements in Greenland from a maritime perspective. In S. M. Lewis-Simpson (ed.), *Vinland revisited: The Norse world at the turn of the first millennium.* Selected Papers from the Viking Millennium International Symposium, Newfoundland and Labrador, Historic Sites Association of Newfoundland and Labrador, St. John's.

———. 2004. The Ormen Lange marine archaeology project. Oceans-Technocean Conference, Kobe, Japan.

———. 2008. Seven seas: Maritime archaeology at the Norwegian University of Science and Technology. In *Collaboration, communication and involvement: Maritime archaeology and education in the 21st century.* Nicolaus Copernicus University, Torun, Poland.

———. 2008. *Ormen Lange deep-water archaeology project, report and catalogue.* Vitenskapsmuseet, NTNU.

Jasinski, M. E., B. Sortland, and F. Søreide. 1995. Applications of remotely controlled equipment in Norwegian marine archaeology. Oceans 1995 Conference Proceedings, San Diego.

Kaijser, I. 1994. Från tvättlina till Sjöuggla—en fotohistoria under vatten. *Sjöhistorisk årsbok* 1994–1995: 155–72.

Kelland, N. C. 1991. Acoustics as an aid to salvage location and recovery. *Hydrographic Journal,* no. 60: 27–33.

Kiernan, M. 1997. Image processing: The sonar revolution. 4th Underwater Science Symposium, Society for Underwater Technology, Newcastle upon Tyne.

Kinder, G. 1995. Quest for the ship of gold. *Reader's Digest,* March: 202–38.

King T. F. 2008. Robotic archaeology on the deep ocean floor. In D. M. Pearsall (ed.), *Encyclopedia of Archaeology.* Academic Press.

Kingsley S. 2003. Odyssey Marine Exploration and deep-sea shipwreck archaeology: The state of the art. *Minerva* 14.3.

L'hour, M. 1993. The wreck of a Danish merchant ship, the Sainte Dorothéa (1693). *International Journal of Nautical Archaeology* 22.4: 305–22.

Long, L. 1995. Les archéologues au bras de fer: nouvelle approche de l'archéologie en eau profonde. In Cent (100) sites historiques d'intérêt commun méditerranéen. Protection du patrimoine archéologique sous-marin en Méditerranée—Marseille: Atelier du patrimoine: 14–46 (Documents techniques, V).

Lore, D. 1990. Cargo delivered. *Sandlapper,* March/April: 10–15.

Ludvigsen, M., H. Singh, et al. 2006. *Photogrammetric models in marine archaeology.* Oceans 2006, Boston, IEEE/MTS.

Ludvigsen, M., and F. Søreide. 2006. Data fusion on the Ormen Lange shipwreck site. Oceans 2006 Conference Proceedings, Boston.

Ludvigsen, M., B. Sortland, et al. 2007. Applications of georeferenced underwater photo mosaics in marine biology and archaeology. *Oceanography* 20(4): 140–49.

MacDonald, I. R., and S. K. Juniper. 1997. Sipping from the fire-hose: How to tame the data flow from ROV operations. *MTS Journal* 31.3: 61–67.

MacDonald, I. R., J. F. Reilly II, J. S. Chu, and D. Olivier. 1997. NR-1: Deep-ocean introduction of new laser line scanner. *Sea Technology* 38, February: 59–64.

Macy, B. D., 1993. Automating complex underwater manipulation. Underwater Intervention conference, New Orleans.

Marx, R. F. 1990. *The underwater dig: Introduction to marine archaeology.* Pisces Books.

———. 1993. *The search for sunken treasure.* Swan-Hill Press.

———. 1996. Treasure hunting goes high-tech. *Silver Kris,* March: 62–66.

Mazel, C. 1985. *Side-scan sonar record interpretation.* Klein Associates.

McCann, A. M. 2000. Amphoras from the deep sea: Ancient shipwrecks between Carthage and Rome. *Rei Cretariæ Romanæ Favtorvm Acta* 36: 443–48.

———. 2001. An early imperial shipwreck in the deep sea off Skerki Bank. *Rei Cretariæ Romanæ Favtorvm Acta* 37: 257–64.

McCann, A. M., and J. Freed J. 1994. Deep water archaeology: A late-Roman ship from Carthage and an ancient trade route near Skerki Bank off north-west Sicily. *Journal of Roman Archaeology,* s.s. no. 13.

McCann, A. M., and J. P. Oleson. 2004. Deep water shipwrecks off Skerki Bank: The 1997 survey. *Journal of Roman Archaeology,* s.s. no. 58.

Mearns, D. L. 1995. Search for the bulk carrier Derbyshire: Unlocking the mystery of bulk carrier shipping disasters. In *Man-made objects on the seafloor—discovery, investigation and recovery.* Society for Underwater Technology, London.

Medard, J. 1997a. New 3D underwater local positioning system dedicated to archaeology and topographical needs. *Underwater Systems Design* 19.3: 5–9.

———. 1997b. Underwater positioning for archaeology and topographical needs. *Sea Technology* 38.7: 31–33.

Mindell, D. A., and B. Bingham. 2001. New archaeological uses of autonomous undersea vehicles. Oceans 2001 Conference Proceedings, Honolulu.

Mindell, D. A., and K. L. Croff. 2002. Deep water, archaeology and technology development. *MTS Journal* 36.3: 13–20.

Mindell, D. A., H. Singh, D. Yoerger, L. Whitcomb, and J. Howland. 2004. Precision mapping and imaging of underwater sites at Skerki Bank using robotic vehicles. In A. M. McCann and J. P. Oleson (eds.), Deep-water Shipwrecks off Skerki Bank: The 1997 Survey. *Journal of Roman Archaeology,* s.s. no. 58.

Momber, G. 2000. Reflections on the legal standing of underwater archaeology in the UK. *Journal of the Society for Underwater Technology* 24.3: 115–18.

Morgan, N. 1990. Ocean environments. In N. Morgan (ed.), *Marine technology reference book.* Butterworths.

Muckelroy, K. 1978. *Maritime archaeology.* Cambridge University Press.

———. 1980. *Archaeology under water: An atlas of the world's submerged sites.* McGraw-Hill.

Murphy, L. E, and A. R. Saltus. 1990. Considerations of remote sensing limitations to submerged historical site survey. Underwater Archaeology Proceedings from the Society for Historical Archaeology Conference, Tucson.

Nævestad D. 1991. *Kulturminner under vann—vurdering av nye tiltak i forvaltningen.* NAVF, Norsk Sjøfartsmuseum.

Nelson, D. A. 1983. Hamilton and Scourge ghost ships of the war of 1812. *National Geographic,* March: 288–313.

Newton, I. 1989. Underwater photogrammetry. In I. Karara (ed.), *Non-topographic photogrammetry.* American Society for Photogrammetry and Remote Sensing.

Noonan, T. 1992a. The greatest treasure ever found. *Life,* March: 32–42.

———. 1992b. History under the sea. *American History Illustrated,* July/August: 58–62.

Papatheodorou, G., M. Geraga, and G. Ferentinos. 2005. The Navarino naval battle site, Greece, an integrated remote-sensing survey and a rational management approach. *International Journal of Nautical Archaeology* 34.1: 95–109.

Peacock, D. P. S., and D. F. Williams. 1986. *Amphorae and the Roman economy.* Longman.

Pearson, C. (ed.). 1987. *Conservation of marine archaeological objects.* Butterworths.

Penvenne, L. J., and J. Penvenne J. 1994. ISIS: Versatile sonar data acquisition. *Sea Technology* 35, June: 59–65.

Pickford, N. 1993. *Atlas of shipwreck and treasure.* Dorling Kindersley.

Pizarro, O., R. Eustice, and H. Singh. 2004. Large area 3D reconstructions from underwater surveys. Oceans 2004 Conference Proceedings, vol. 2, Kobe, Japan.

Pizarro, O., and H. Singh. 2003. Towards large area mosaicing for underwater scientific applications. *IEEE Journal of Oceanic Engineering* 28.4: 651–672 (Special Issue on Underwater Image and Video Processing).

Plets, R. M. K., J. K. Dix, and A. I. Best. 2008. Mapping of the buried Yarmouth Roads Wreck, Isle of Wight, using a Chirp sub-bottom profiler. *International Journal of Nautical Archaeology* 37.2: 360–73.

Quinn, R., J. R. Adams, J. K. Dix, and J. M. Bull. 1998. The Invincible (1758) site: An integrated geophysical assessment. *International Journal of Nautical Archaeology* 27.2: 126–138.

Quinn, R., J. M. Bull, and J. K. Dix. 1996. Shipwreck surveying and CHIRP (sub-bottom profiler) technology. 3rd Underwater Science Symposium, Society for Underwater Technology, Bristol, U.K.

———. 1997a. Buried scour marks as indicators of palaeo-current direction at the Mary Rose wreck site. *Marine Geology* 140: 405–13.

———. 1997b. Imaging wooden artefacts using CHIRP sources. *Archaeological Prospection* 4: 25–35.

Quinn, R., J. M. Bull, J. K. Dix, and J. R. Adams. 1997. The Mary Rose site: Geophysical evidence for palaeo-scour marks. *International Journal of Nautical Archaeology* 26.1: 3–16.

Quinn, R., M. Dean, M. Lawrence, S. Liscoe, and D. Boland. 2005. Backscatter responses and resolution considerations in archaeological side-scan sonar surveys: A control experiment. *Journal of Archaeological Science* 32: 1252–64.

Richards, J. D., and N. S. Ryan. 1985. *Data processing in archaeology.* Cambridge University Press.

Robinson, W. S. 1981. *First aid for marine finds.* Handbooks in maritime archaeology, no. 2. National Maritime Museum, London.

Rokoengen, K., and A. B. Johansen. 1996. Possibilities for early settlement on the Norwegian continental shelf. *Norsk Geologisk Tidsskrift* 76: 121–25.

Rønnby, J. 1995. Activity, degradation or natural sedimentation. In I. Vuorela (ed.), *Scientific methods in underwater archaeology.* Council of Europe, Strasbourg.

Rule, N. 1989. The direct survey method (DSM) of underwater survey, and its application underwater. *International Journal of Nautical Archaeology and Underwater Exploration* 18.2: 157–62.

———. 1995. Some techniques for cost-effective three-dimensional mapping of underwater sites. From the Computer Applications and Quantitative Methods in Archaeology conference, 1993, Staffordshire University. *BAR International Series* 598: 51–56.

Ryther, J. H., J. P. Fish, and D. B. Harris. 1991. Investigating the application of using remotely operated vehicles (ROV) for shipwreck investigations. *ROV Intervention* 91: 192–199.

Saade, E. J., and D. Carey. 1996. Laser-line scanner: High-grading search targets. *Sea Technology* 37, October: 63–65.

Sewell, M. 1994. Principles of underwater lighting systems. *Underwater Systems Design* 16.2: 5–15.

Shomette, D. 1987. The Pitcher wreck: An exercise in crisis. Underwater Archaeology Proceedings from the Society for Historical Archaeology Conference, Savannah, Ga.

———. 1988. The New Jersey project: Robots and ultrasonics in underwater archaeological survey. Underwater Archaeology Proceedings from the Society for Historical Archaeology Conference, Reno, Nev.

Singh, H., J. Adams, B. P. Foley, and D. Mindell. 2000. Imaging for underwater archaeology. *American Journal of Field Archaeology* 27.3: 319–28.

Singh, H., J. Howland, and O. Pizarro. 2004. Large area photomosaicking underwater. *IEEE Journal of Oceanic Engineering* 29.3: 872–86.

Singh, H., L. Whitcomb, D. Yoerger, and O. Pizarro. 2000. Microbathymetric mapping from underwater vehicles in the deep ocean. *Journal of Computer Vision and Image Understanding* 79.1: 143–61.

Snowball, G. 1996. Stabilising air bag buoyancy. *Underwater Systems Design* 18.6: 4–5.

Sognnes, K. 1985. King Øystein's harbour at Agdenes, Norway. Conference on Water Front Archaeology, North-European Towns, no. 2, Historisk Museum, Bergen.

Søreide, F. 1999. *Applications of underwater technology in deep water archaeology: Principles and practice.* Norwegian University of Science and Technology.

———. 2000. Cost-effective deep water archaeology: Preliminary investigations in Trondheim Harbour. *International Journal of Nautical Archaeology* 29.2: 284–93.

Søreide, F., and A. D. Atauz. 2002. Deepwater: The future of marine archaeology? Some examples from the Mediterranean. *Marine Technology Society Journal* 36.3: 21–32.

Søreide, F., S. Høseggen, M. E. Jasinski, S. Kristiansen, and B. Sortland. 1996. Information processing in marine archaeology. Oceans 1996 Conference Proceedings, Ft. Lauderdale, Fla.

Søreide, F., and M. E. Jasinski. 1998. The Unicorn wreck, central Norway: Underwater archaeological investigations of an 18th-century Russian pink, using remotely controlled equipment. *International Journal of Nautical Archaeology* 27.2: 95–112.

———. 2000. Marine archaeology and protection of heritage in deeper water: Consequences for future offshore construction projects. Oceans 2000 Conference Proceedings, Providence, R.I.

———. 2005. Ormen Lange: Investigation and excavation of a shipwreck in 170m depth. Oceans 2005 Conference Proceedings, Washington, D.C.

———. 2008. Ormen Lange, Norway: The deepest dig. *International Journal of Nautical Archaeology* 37.2.

Søreide, F., M. E. Jasinski, and T. O. Sperre. 2006. Unique new technology enables archaeology in the deep sea. *Sea Technology* 47.10: 10–13.

Søreide, F., S. Kristiansen, and M. E. Jasinski. 1997. VETIS: A survey tool for marine archaeology. 4th Underwater Science Symposium, Society for Underwater Technology, Newcastle upon Tyne.

Sprunk, H. J., P. J. Auster, L. L. Stewart, D. A. Lovalvo, and D. H. Good. 1992. Modifications to low-cost remotely operated vehicles for scientific sampling. *MTS Journal* 26.4: 54–58.

Stangroom, J. E. 1995. Subsea manipulation of heavy loads using fresh water: A concept study. *Underwater Technology* 21.2: 30–40.

Steffy, J. R. 1994. *Wooden ship building and the interpretation of shipwrecks.* Texas A&M University Press.

Stemm, G. 1992. The future of deep water: Commercial archaeology, a new business? *Treasure Diver* 4.2: 28–33.

Stewart, W. K. 1990. A model based approach to 3D imaging and mapping underwater. *Journal of Offshore Mechanics and Arctic Engineering* 112, November: 352–56.

———. 1991. Multisensor visualization for underwater archaeology. *IEEE Computer Graphics and Applications,* March: 13–18.

Stone, L. D. 1992. Search for the SS Central America: Mathematical treasure hunting. *Interfaces* 22:1: 32–54.

Størkersen, N., and A. Indreeide. 1997. Hugin: An untethered underwater vehicle system for cost effective seabed surveying in deep waters. Underwater Technology International Conference, Aberdeen.

Stowe, K. 1996. *Exploring ocean science.* John Wiley and Sons.

Thanem, R. W. 1970. Hovmod kostet 700 russiske emigranter livet. *Adresseavisen,* July 1970.

Throckmorton, P. (ed.). 1991. *The sea remembers: Shipwrecks and archaeology.* Smithmark.

Thurman, H. V. 2003. *Introductory oceanography.* Prentice Hall.

Tolson, H. 2005. A 19th century shipwreck and the myth of deep ocean preservation. Oceans 2005 Conference Proceedings, Washington, D.C.

Tusting, R. F. 1996. Lasers underwater. *Underwater Systems Design,* March/April.

Tyler, P. A., A. L. Rice, C. M. Young, and A. Gebruk. 1996. A walk on the deep side: Animals in the deep sea. In C. P. Summerhayes and S. A. Thorpe, *Oceanography: An illustrated guide.* Manson Publishing.

Varmer, O., and C. M. Blanco. 1999. United States of America. In Sarah Dromgoole (ed.), *Legal Protection of the Underwater Cultural Heritage: National and International Perspectives.* Kluwer Law International.

Ward, C., and R. D. Ballard. 2004. Deep-water archaeological survey in the Black Sea: 2000 season. *International Journal of Nautical Archaeology* 33.1: 2–13.

Warren, D. J., R. A. Church, and K. L. Eslinger. 2007. Deepwater archaeology with autonomous underwater vehicle technology. Offshore Technology Conference 2007, Houston.

Warrinder, P. 1995. Diving techniques or diving technology. 2nd Underwater Science Symposium, Society for Underwater Technology, University of Stirling.

Watts, G. P., Jr. 1987. A decade of research: Investigation of the USS Monitor. Underwater Archaeology Proceedings from the Society for Historical Archaeology Conference, Savannah, Ga.

Webb, D. L. 1993. Seabed and sub-seabed mapping using a parametric system. *Hydrographic Journal,* no. 68: 5–13.

Webster, S., O. Pizarro, and H. Singh. 2001. Photomosaicking in underwater archaeology. *INA Quarterly* 28.3: 22–26.

Westenberg, B. 1995. En fjärrstyrd robot i marinarkeologiens tjänst—Söugglan. Marinarkeologi—kunnskapsbehov. Rapport fra seminar, Norges forskningsråd, Korshavn ved Lindesnes.

Whitcomb, L. L. 2000. Underwater robotics: Out of the research laboratory and into the field. IEEE 2000 International Conference on Robotics and Automation, San Fransisco.

Wickler, S., M. E. Jasinski, F. Søreide, and O. Grøn. 1999. Remote sensing in marine archaeology: Preliminary results from the Snow White project, Arctic Norway. *World Archaeological Congress 4.*

Wilkinson, A. (ed.). 1979. *The oceans.* Grisewood and Dempsey.

Wright, A. C. 1997. Deep-towed sidescan sonars. *Sea Technology* 38, June: 31–38.

Zarzynski, J. W., K. B. McMahan, B. Benway, and V. J. Capone. 1995. The 1758 Land Tortoise Radeau shipwreck: Creating a seamless photomosaic using off the shelf technology. Underwater Archaeology Proceedings from the Society for Historical Archaeology Conference, Washington, D.C.

INTERNET RESOURCES

America's Lost Treasure, the S.S. *Central America*,
 http://www.sscentralamerica.com/
Deep Water Archaeology Research Group,
 http://web.mit.edu/deeparch/
Deep Wrecks Project, PAST Foundation,
 http://www.pastfoundation.org/DeepWrecks/
Institute of Nautical Archaeology,
 http://inadiscover.com
JASON Science,
 http://www.jason.org
Mardi Gras Shipwreck,
 http://www.flpublicarchaeology.org/mardigras/
Minerals Management Service, U.S. Department of the Interior,
 http://www.mms.gov/
Monitor National Marine Sanctuary,
 http://monitor.noaa.gov/
Mystic Aquarium, Institute for Exploration,
 http://www.ife.org
Nauticos Corporation,
 http://www.nauticos.com/ancientwreck.htm
NOAA Office of Coast Survey,
 http://www.nauticalcharts.noaa.gov
Nordic Underwater Archaeology,
 http://www.abc.se/~m10354/uwa/
Norwegian University of Science and Technology,
 http://www.hf.ntnu.no/maritime/
Odyssey Marine Exploration,
 http://www.shipwreck.net/
Project Archeomar, Ministero per i Beni e le Attività Culturali,
 http://www.archeomar.it/
ProMare, http://www.promare.org/
Société d'Etudes en Archéologie Subaquatique,
 http://www.archeo-seas.org/
Underwater Archeology Exhibition, Ministère de la culture et de la communication,
 http://www.culture.gouv.fr/culture/archeosm/en/
United Kingdom Hydrographic Office,
 http://www.ukho.gov.uk
University of Rhode Island, Graduate School of Oceanography,
 http://www.gso.uri.edu/
Woods Hole Oceanographic Institution,
 http://www.whoi.edu/

INDEX

Page numbers for illustrations are in *italic*.

3H Consulting Ltd. (company), 31, 153
Academy of Science (Russia), 64
Aegaeo (Greek research vessel), 51
Aegean Sea, 19, 45, 46, 51, 53, 164
Agdenes, 56, *57*
Agios Athanasios, 48
Agios Petros (bay of), 46
Akademik (Bulgarian research vessel), 54
Alcoa Seaprobe (research vessel), 23–25
ALL-NAV (navigation system), 25
amphora (location site and period), *49*, 54, 156; Aegean Sea, 51; "Alonnesos wreck," 46, *47*, 47, 48; *Arles IV*, 32–33, *33*; Basses de Can, 34; Black Sea, 51; Bulgarian coast, 54–55; Cap Bénat, 33; Dry Tortugas, 84–85; Egyptian coast, 19; "Elissa" wreck (Phoenician), 44; "Grand Ribaud F" wreck (Etruscan), 35; Greek Sporades (Mendian), 46; *Héliopolis 2*, 31; Ionian Sea (Roman), 49; *Isis*, 37; Lipari (island), 35–36; Mediterranean (Rhodian), 98; "Melkarth" wreck, 93; Ormen Lange (Roman), 145; *Sud-Caveux*, 34; "Tanit" wreck, 44; Tyrrhenian Sea, 23; Xlendi harbor, 40, 42–43, *42*; Vasiliko Bay (Byzantine), *46*, 47, 48. See also artifacts
Anglo-American War of 1812, 27
Antarctica, 159
AOSC (company, Aberdeen, Scotland) 85
AQUA-METRE D100 (positioning system), 118–19
archaeological oceanography (defined), 7; programs in, 52
Archeomar project, 39

Archéonaute (French research vessel), 31, 35
Arctic Discoverer (research vessel), 95
Arctic Ocean, 4, *5*, 159
Argo (towed survey system), 19, 37
Arkhangelsk, Russia, 62–63
artifact handling: conservation of, 10, 31, 88, 133, 145, 149, 150, 154; deterioration/preservation, 158–61; removal, 3; risks from biological processes, 162; risks from physical forces, 161–62; sampling, 15, 115, 132, 139—examples of, 36, 42, 83, 90
artifacts: amphora (*see main entry* amphora); anchors, 29, 44, 49, *50*, 57, 62, 93, 98—of iron, 19, 32, 36, 37, 49—stocks for, 36, 98—of stone, 44; bells, 81–82, 85, 95; brass/bronze, 73—bell, 85—helmet, 51—spear point, 51;—telescope, 87; cannon, 23, 34, 35, 29, 61, 62, 63, 64, 65, *65*, 87, 89, 110, 115, 145—cannonballs, 87; ceramic, 29, 40, 52, 60, 61, *80*, 81, *81*, 85, 86, 87, 109, 145, 155, 158; coins, 35, 65, *65*, 81, 83, *83*, 89–90, 92, *92*, 93, 94–97, 145, 150 (*see also* artifacts: gold; silver); copper, 145—clad, 19, 28, *28*, 91—ingots, 32—nails, 36—pots, 87; glass, 29, 61, 65, 72, 73, 87, 92, 145, 146, 155, 158, 162; gold, 81, 87, 89, 92, *92*, 93, 95–96, *96*; iron, 36, 55, 61, 65, 68, 70, 72, 73, 93 (*see also* anchors); lead, 36, 60, 61, 62, *64*, 65, 66, 68, 70, 72, 87, 145; miscellaneous, 87:—arrows, 150—bricks (yellow), 60,—buttons, 81—camboose (ship's stove), 29—dagger handles, 150—gun flints, 65—nautical instruments, 29—personal belongings, 96—seeds, 81, 87;

pewter, 73, 145; porcelain, 29, 65, 73, 92; pottery, 19, 32, 92, 145—Athenian, 46—Campanian, 35–36;—shards, 52, 86—stoneware, 72, 73; silver, 87, 89, 93, 145; stone(s), 145,—anchor, 44—millstone, 37—ballast, 44, 85, 86, 87, 156; textiles, 96, 145; wood, 61, *64*, 66, 68, 107;—bulkhead frame, 57;—chest, 29;—gauiacum wood, 65, *69*, 70;—gun carriage, 65;—hull, 28, 29;—mast, 52;—nails, 52, 56;—planking, 37, 55, 56, 59;—structural timber, 72–73, 77, 85, 93
Asteris. *See* Daskalio islet
Atlantic Ocean, 4, *5*, 19, 93, 164; off European coast, 31, 43; off U.S. coast, 91, 92
AUVs (autonomous underwater vehicles), 13, 20–21; camera-mounted, 110; Hugin, *20*, 28; for mapping, 103; Maridan, 21; SeaBED, 51
Azores, 4

bathymetry, 76, 108, 110; bathymeric maps, 25, 44, 51, 74, 89, 90, 103, 133, 141; microbathymetry, 21, 39, *82*, *125*; multibeam, 30, 125–26
Bailey & Leetham (company), 63
Balearic Islands, 32, 43
Ballard, Robert D., 3, 118, 145, 151; and Black Sea project, 52, 54; and the Jason project, 36–37; and Sea of Crete project, 51; and the *Titanic*, 36, 156; and WHOI, 19, 36, 43
Baltic Sea, 55–56, 127, 156, 160, 164
Bascom, Willard, 5, 23, 24–25, 156
Basilicata (region of Italy), 39
Bay of Cannes, 35
Benthos (company), 27

177

Bétique, Spain, 32
BioScan (company), 26
Birkenhead (England), 35
Bjørkvik, Eilert, 63, 64
Black Sea, 46, 52–55, 127, 164
Boeing 727 (wreck of), 25
Bosphorus Strait, 52, 54, 164
BP Exploration, 28
Brazil, 165
Buhl, Theodore DeLong, 28
Bulgakov, Vasilij, 63
Bulgaria, 54
Bulgarian Academy of Sciences, 55
Bulgarian Center for Underwater Archaeology, 54
Byzantine period, 52; Early, 46; Middle, 46, 47; twelfth century, 48, 49. *See also* shipwreck; Vasiliko Bay (Byzantine) site
Byzantium, 46

C&C Technologies, 21; Hugin, *20*, 28
Caesarea, Israel, 110
Calabria (region of Italy), 39
cameras: acoustic, 132; CCD (charge coupled devices), 121; digital, 21, 120, 121, 125; HD (high-definition), 74, 121; ICCD, 121; ISIT, 121; laser line scan, 112, 115, 129, 132; SIT (silicon intensifier target), 36, 37, 85, 121; and site illumination, 95, 122; and stereoscopic photography, 25, 32–34, 85, 122; SVHS video recorder, 85; 3CCD, 76, 85, 121; towed, 110–12; use of, 13–15, 20–21, 75–76, 140, 141, 151; video, 14, 20, 111, 112, 119, 120, 126, 129. *See also* photogrammetry; photomosaic technology (photomosaics)
Campania (region of Italy), 39
Campos y Pineda, Gabriel, de 96
Canada, 129, 164
Canadian Archaeological Institute (Athens), 51
Cap de Saint-Tropez, France, 34
Cape Hatteras, 19, 25
Carolus III (Bourbon king of Spain), 96
Carolyn Chouest (research vessel), 28, 37
Carthage, 37, 40, 43
Cascais, Portugal, 55
Catherine the Great (of Russia), 55
Cefalonia Piccola. *See* Ithaki(Ionian island)
Cehili (research vessel), 74, *75*
Center of Scientific Research (France). *See* CNRS

Central Navy Museum, (of Russia, St. Petersburg), 63, 64
Chaniotis, Dimitris, 46
Cherbourg (France), 35
Chersonesos (Greek colony), 53
Chios (Greek island), 46, 51
CNANS (Portuguese National Center for Nautical and Underwater Archaeology, 55
CNRS (Center of Scientific Research, France), 31, 35
CodaOctopus (company), 132
Colombia, 165
COMEX Industries, 31, 33, 34, 35
Conoco (company), 61, 167
Conservation Research Laboratory, 31
Coopernaut Franca (research vessel), 39
Corsica, 43
Cousteau, Jacques-Yves, 31
Crete, 45. *See also* Sea of Crete
Crimean Peninsula, 53, 54
Crumlin-Pedersen, O., 5
Cuba, 84, 96
Cutty Sark (clipper ship), 55
Czech Republic, 165

Darius (Persian king), 51
Daskalio islet (Ionian Sea), 48
data archiving and analysis (software), 26, 133–35; ArcGIS (ESRI), 76, 131, 153; ArcMap (ESRI), 81, 153; AutoCad, 126, 136; CAD, 119, 128, 129, 136, 153; CODA (data format), 131; DirectShow API (Microsoft), 153; ESRI (company), 39, 81, 153; Optimas (BioScan), 26; PhotoModeler (Windows), 129–30; SEG-Y (data format), 131; SES (data format), 131; Site Surveyor, 126; VisualEvent, 122; VisualSoft, 153; VisualWorks, 122
Deane, (brothers Charles and John), 3
Deep Ocean Engineering, 85
Deep Sea Systems International, 26
Department for Underwater Archaeological Research, France. *See* DRASSM (Department for Underwater Archaeological Research, France)
DGPS (differential global positioning), 45, 57, 59, 66, 103. *See also* DP (dynamic positioning): GPS
Dictionary of Disasters at Sea during the Age of Steam, 4
Digiquartz depth gauge, 86
diving: using air, 10, 11; using armored suits, 18; using atmospheric system, 18—WASP, 97; using bells, 3, 12, 13, 36; and decompression, 10, 12, 13, 18; and bounce diving technique, 12, 13; using helmets, 3, *4;* using mixed-gas, 12; using rebreathers, 12; suits, 3, 12; and saturation diving, 12, 13; and scuba equipment, 3, *4,* 6, 12, 13, 26, 51, 64, 83, 155
Dons, Henrik, 64
Doppler (shift), 21, 44, 76, 117
DP (dynamic positioning) systems, 23, 72, 88, 90, 95, 119. *See also* DGPS (differential global positioning): GPS (global positioning)
DRASSM (Department for Underwater Archaeological Research, France), 31–35
Dry Tortugas (Florida keys), 84, 87

Easington, U.K., 70
Eastport International (company), 25
Echoscope, 132
echosounder, 102, 103, *104*, 132, 133
Edgetech (company); 91, 131
Edwin Link (research vessel), 87
EEA (Greek Ephorate of Underwater Antiquities), 45, 46, 51
Egypt, 43, 45; and battle of Navarino Bay, 52; coast of, 19
Elba (island), 23
Electronic Data Systems, 36
Ellen (Norwegian bark), 93
Endeavor (research vessel), 51, 53
ESRI (geographic information system software company), 39, 81, 153
Europipe (gas pipeline), 168
EXACT (precision acoustic navigation system), 36, 37, 44, 117–18
Exogi (cape of Ithaca), 48
excavation, 7–8, 9–10; and artifact retrieval, 144–46, 148–50; and complete site removal, 148–50; documentation and recording of, 146, 150–51; and organic materials, 145; and postprocessing, 154; and remote intervention systems, 139–40 (*see also* ROVs); and risk to the site, 139; and sediment removal, 139, 141, 143; surveys, 126, 133; virtual, 32–34, 131. *See also* excavation equipment
excavation equipment: airlift suction devices, 61, 143, 149, 150; and balloon lifters, 146, 148, 150; and collection devices, 14, 61, 81, 91, 143, 146, 150; dredges, 141, *142,* 143; thrusters, 119,

120, 139, 141, 148; manipulator devices, 14, 15, 132, 134, 139–46, 148—Schilling, 70, 88, 90; suction pickers, 29, *30,* 81, 95, 145; vacuum devices, 95—SeaVac, 95; water jets, 24, 52, 95, 141, 143, 149. *See also* excavation
Exploramar (expedition company), 96

Ferranti Track Point II (system), 85
Fiscardo, Kefalonia, 49
Florida State University, 12
Flow, Jenette, 171
France, 43, 81, 84, 165; and deepwater archaeology, 8, 17, 31; and the *Epave aux Ardoises*, 119. *See individual governmental offices by name*
Franchthi cave (Peloponnesos), 45

Gagnan, Émile, 31
Garrison, E. G., 6
General Dynamics, 17, 18
General Motors, 17
George Lawley and Sons (Boston, shipbuilders), 28
Georgia (coast of), 91
Germany, 165
Gibraltar, 21, 89
GIS (geographic information) system, 49, 119, 133; and Archeomar project, 39; and ArcGIS, 76, 131, 153; and mapmaking system, 125, 134; from 3H Consulting, 153; and WebGIS, 39
Gotland (Swedish island), 156
GPS (global positioning); using buoys, 21; defined, 101–102, *102;* ship positioning, 49, 103, 153; and site mapping, 101; and site records, 25, 36, 39; and target sites, 119
Grand Ribaud Island (France), 35
Great Lakes (U.S.), 18, 27, 164
Greece, 21, 45, 46, 51, 55; and UNESCO convention on underwater cultural heritage, 165. *See also* Ionian Islands
Greek Ephorate of Underwater Antiquities. *See* EEA
Greek-Norwegian Deep-Water Archaeological Survey, 45
Grumpy Partnership, 96–97
Guinea-Bissau, 165
Gulf of Lion, 32
Gulf of Mexico, 18, 19, 21; deepwater archaeology in, 27–31; Mississippi Canyon of, 19, 28
Gulf of Morbihan (Brittany), 119

Hadjidaki, Elpida, 46, 47
Håkonson, Håkon (king of Norway), 57
Haltenbanken (in Norwegian Sea), 61, 167
Haltenpipe Development Project (HDP), 61, *66, 68,* 70, 167, 168
Hamilton, Donny, 31
Hamilton-Scourge Foundation/National Geographic Society, 27. *See also* shipwrecks; *Hamilton* (schooner); *Scourge* (schooner)
Harbor Branch Foundation, 25
Harbor Branch Oceanographic Institution, 87
Havana (Cuba) 84, 96
HCMR. *See* Hellenic Center for Marine Research
HDP. *See* Haltenpipe Development Project (HDP)
Heidrun oil and gas field, 61, 167
Hellenic Center for Marine Research (HCMR), 51
Herodotus, 51
Hitra (island), Norway, 61, 62 167
HMS *Belos* (Swedish Naval rescue ship), 55
Homer, 48
Hustadvika (Norwegian coastline), 72

Iceland, 165
IFE. *See* Institute for Exploration
IFREMER (French Research Institution for the Exploitation of the Sea), 31–32
Imagenex (Technology), 39
INA (Institute of Nautical Archaeology), 35, 39, 98
Indian Ocean, 4, *6,* 159, 164
Innomar (technology company), 131
Institute for Exploration (IFE), 43, 51, 52
Institute of Nautical Archaeology. *See* INA
Ionian Islands, 48
Ionian Sea, 19, 48–49
Israel, 43–44, 165
Italy, 19, 36, 43, 49
Ithaki (Ionian island), 45, 48, 49

Janus (research vessel), 39
Jason Foundation for Education, 37
Jason Project, 36–39
John Lair and Sons shipyards, 35

Kalman filter (algorithm), 117
Kea island (Greek Cyclades), 19
Kefalonia (Ionian island), 48, 49

King Herod, 110
King Øystein harbor (Norway), 56–57
King Sigurd (of Norway), 57
Klein (Associates) 59, 84
Kongsberg (company), 117; and HAINS navigation system, 30; and scanning sonar, 27, 59, 66, 74, 85, 95; and SSBL system, 29–30, 49, 59, 61, 66; and TOPAS, 60
Kootenay Lake, British Columbia, 27
Kos (Greek isle), 98
Kraft (TeleRobotics), 81
Kritzas, Charalambos, 47
Kronstad, Russia, 62
Kyra-Panagia (island), 46

Lake Ontario, 27
LBL (long baseline underwater positioning system): accuracy of, 115, 117, 127; for locating, 33; for positioning, 37, 76, 90, 95; for provenience recording, 81, 86; technology of, 115
Lilybaeum, Sicily, 43
Liverpool, 29
Lloyds' Register, 4
Loran C (navigation system), 84
Louisiana: coast of, 29; Department of Culture, Recreation, and Tourism: Division of Archaeology, 31

Macedonia, 45
magnetometry, 39, 108–109; cesium magnetometers, 110; Geometrics G822 magnetometers 132; and magnetic anomalies, 110; nuclear resonance magnetometer, 109; and proton system, 109–110; use of Seaspy Overhauser technology, 91; surveys using, 68, 69, 70, 84, 110, 132; Varian proton magnetometer, 25
Magnus I (king of Norway), 57
mapping. *See* bathymetry
Malta, 39–40, 43
Marex International (Memphis, Tenn.), 97
Mariana Trench, 17, 159
Marine (brig), 93
Marine Sonics, 45
Marx, R. F., 84
Massalia (Greek port), 46
McCann, Anna Marguerite, 3
Medea (towed support vehicle), 36, 44
Mediterranean Sea, 5, 19, 21, 23, 43, 164; eastern, 43–52; western, 31–43
Melos, 45

Mendi (Greece), 46
Microsoft (software company), 122, 153
Minibex (research vessel), *34*, 35, 39
Ministry for Higher Education and Research (France), 31
Ministry of Cultural Heritage (Italy), 39
Ministry of Culture (France), 31
Ministry of Culture (Greece), 45, 47
Mistake (fishing trawler), 96
MIT (Massachuetts Institute of Technology), 51, 118, 131
MMS. *See* United States; Department of Interior: Minerals Management Services (MMS)
Molde, Norway, 72
Moscow, 65
Mt. Athos, 51
Muckelroy, K., 10
Munkholmen (island in Trondheim harbor), 59
Murphy, Jerry, 96
Museum of Archaeology, Lisbon, 4
Mystic Aquarium (Connecticut), 43

Nævestad, D., 56
Nantucket Island, 19
National Geographic Society, 27, 36
National Museum of Denmark: Centre for Maritime Archaeology, 21
National Science Teachers Association, 37
Native American deep water archaeological sites, 27
Nautical Archaeology Associates, 26
Nauticos Corporation, 97–98
Nautilion (expedition), 32
Naval Museum (Russian). *See* Central Navy Museum, (of Russia, St. Petersburg)
Navarino Bay (Pylos), 52
Netherlands, 165
New Orleans, 29, 91, 96
New York, 91, 93
Nigel Boston (company), 150
NOAA, 19, 26; Office of Coast Survey, 4
Nord Stream (gas pipeline), 56
Norsk Hydro, 70; pipeline proposal, 70, 72, 73
North Sea, 6, 62, 164, 167
Northern Maritime Research: Northern Shipwrecks Database, 4
Northern Sporades (Greek archipelago), 45
Norway, 3, 70, 165; coastline of, 4, 6, 56; and deep water archaeology, 8, 17, 55, 61–62, 81, 140, 167, 168. *See also* NTNU; Orem Lange (pipeline) project
Norwegian Cultural Heritage Act, 70, 84, 166–68
Norwegian Directorate for Cultural Heritage, 72
Norwegian Institute at Athens (NIA), 45
Norwegian Petroleum Directorate, 168
Norwegian Sea. *See* Haltenbanken; *see also* Ormen Lange (pipeline) project
NTNU (Norwegian University of Science and Technology): development and use of deepwater technologies, 3, 40, 45, 55, 59, 65, 131; Institute of Archaeology, 56, 57, 62; Institute of Marine Technology, 56; and the *Jedinorog* project, 65; and the Cultural Heritage Act, 167–68; and Ormen Lange (pipeline) project, 39, 72, 74; and Statoil (HDP project), 62, 167, 168; and Trondheim Harbor project, 56–57

Oceaneering International, 23, 97
OCS (Outer Continental Shelf), 27
Odysseus, 48
Odyssey Explorer (Class II DP ship), 88, 90
Odyssey Marine Exploration, 87–89, 117; and the *Black Swan*, 93; and the "Blue China" wreck, 92–93; exploration criteria, 88, 93; and the SS *Central America*, 93–96, 143; and the SS *Republic*, 91–92; and the HMS *Sussex*, 89–91, 93, 166
Oinoussia (Greek island), 51
Olav Haraldsson, (king of Norway), 57
Olav Tryggvason, (king of Norway), 57
OmniTech (Norwegian company), 132
On-Line Data Storage and Logging System (Seahawk), 86
Ormen Lange (pipeline) project, 29, 70–84, 117, 168; data collection/analysis, 131–32; excavation process, 78–79, 141, *142*, 143–45, *143*, *147*, 151; use of magnetometers, 132; microbathymetry of, *82*; use of multibeam echosounder, *104*, *105*; shipwreck discoveries of, 81–83; software data recording, 153
Ovsyannikov, Oleg V., 64
Øystein (king of Norway), 57

Pacific Ocean, 4, 6, 17, 159, 162, 163
Paraguay, 165
Peparethos (Greek island), 46
Perry Triton. *See* ROVs *by name:* Triton (Perry) XLS-17
Peter the Great (tsar of Russia), 62
photogrammetry, 25, 32–33, 35, 61, 127–30, 150. *See also* photographic documentation
photographic documentation, 119–21; and backscatter, 27, 107, 112, 120–21, 125, 132; guidelines for, 136–38. *See also* cameras
photomosaic technology (photomosaics), 21, 51, 89, 123, *123*, *124*, 125, 128, 132; of the Dry Tortugas site, 87; of *Hamilton* and *Scrooge* site, 27; of the *Isis*, 39; of the *Mardi Gras*, 30; of the *Monitor*, 25, 26; of the Ormen Lange project, 76; and orthogonal photography, 30, 36; of the Phoenician site, 44; of the *Republic*, 91; of the *Sussex*, 90. *See also* photographic documentation
Portsmouth harbor (England), 3
ProMare, Inc., 19, 39, 49, 55
Puglia (region of Italy), 39
Punic Wars, 40, 43
Pylos. *See* Navarino Bay

Ramsøy fjord, 61, 62, 64, 167
Rath & Raundrap (company), 63
Rebikoff, Dimitri, 13
Red Sea, 160
Rickover, Hyman G., 96
Rockwell (International), 17
Rome, 40, 43
ROTs (remotely operated tools), 60, 61; for excavation tasks, 139–40, *141*, 150
ROVs (remotely operated vehicles), 3, 7, 13–14, 20, 21, 35, 61, 83; for excavation tasks, 139–40, *141*, 146; observation-class, 14–15, *15*, 55, 66; platform mounted, 78, 81, 139; work-class, 15, 17, 29, 61, 69, 139. *See also* ROVs *by name*
ROVs *by name:* Achilles, 51; Aegaeo, 51; Argus, 51, 52, *53*, 61; COMEX-Pro, 34; Deep Drone system, 25; Hercules, 51, 52, *53*, 54; Hyball, 59, 61, 66, 68, 70; Jason, 19, 27, 36–39, *38*, 44; Lagune, 35; Magellan, 23; Merlin, 85, 86, *86*; Mini Range, 26; Nemo, 94, 95, *95*, 96; Phantom, 55—Phantom DHD2, 85—Phantom 300, 85—Phantom 500, 85; Poodle, 13; RPV, 27; SeaOwl, 55, 56, 66; SOLO, 67, 69; Sprint 101, 65, 66; Sub-Fighter 30K, 74; Sub-Fighter 7500, 29; Super Achille class, 35, 39; Triton

(Perry) *XLS-17*, 29–30, *30*; *Zeus*, 88, 89, 90
Russia: artifacts (Russian), 65, *65*, 83 *83*,; and deepwater archaeological activities, 23, 55, 164; and naval activities, 52; and Nord Stream gas pipeline route, 56; Russian Federation, 165; and submersibles, 17, 23. See also shipwreck; *Jedinorog: Voronov*

Sæbu Island, 62, 64; Russian Neck (grounding site of *Voronov*), 62
Samos (Greek island), 51
Santa Monica Bay, California, 25
Sardinia, 43
SBL (short baseline underwater), 117. See also SSBL
Sea of Crete project, 51–52
SeaEye (company), 45
Seahawk (research vessel), 84, 85
Seahawk Deep Ocean Technology (company), 84–87. See also On-Line Data Storage and Logging System (Seahawk)
Seahawk Retriever (research vessel), 85, 86
Seaprobe, see *Alcoa Seaprobe*
Seaquest (company), 84
Seatrac (navigation system), 84, 85
Seaway Commander (diving vessel), *67*, 69
seismic profiling, 68, 112–13, 130, 131
Schilling (company), 70, 88, 90
SHARPS (Sonic High Accuracy Ranging and Positioning System), 117
Shell Oil, 28
shipwrecks: USS *Akron* (ZRS-4 airship), 19; CSS *Alabama*, 35; "Alonnesos wreck," 46, 47, 48; *Alvin* (submersible), 162; *Andrea Doria*, 19; *Anona* (steam yacht), 28; *Ark Royal*, 21; *Arles IV*, 32–33, 34—and virtual excavation, 32; *Atocha* (Spanish) 84; Basses de Can site (Roman), 34; *Batéguier*, 35; *Black Swan*, 93; "Blue China Wreck" (merchant ship), 92–93; HMS *Britannic*, 19; *Cap Bénat* (Roman), 33; SS *Central America* (side-wheel steamer), 93–96, 143; "Chersonesos A" (Byzantine), 54; *City of Ainsworth* (sternwheeler), 27; *Dakar* (Israelite submarine), 19, 97; *Den Waagende Thrane* (Waking Trane), 59, 60–61; *Dzherzynsky* (warship), 54; SS *Egypt*, 18; *Ekaterina* (pre-Dreadnaught warship), 54; *El Cazador* (Spanish brig), 96–97; "Elissa," 44; *Epave aux Ardoises*, 119; Gozo, Malta sites, 39–40 (see also shipwreck; Xlendi site); "Grand Ribaud F" (Etruscan), 35; Hellenistic (Nauticos Corporation site), 97–98; CSS *H. L. Hunley* (submarine), 150; *Hamilton* (schooner), 27; *Héliopolis 2* (near Toulon), 31; *Invincible* (18th century), 130; *Isis* (Roman) 37, 38, 39; *Jedinorog* (Russian "pink"), 60–65, 70; *Kronan* (royal ship), 55; *Lenin* (warship), 54; Lipari (third century BC sites), 35; USS *Maine* (second-class battleship), 96; *Mardi Gras*, 29–31, *30*; *Mary Rose* (16th century carvel), 94, 130–31, 149–50; "Melkarth" (Punic or Phoenician), 93; USS *Monitor*, 19, 25–26; *New Jersey* (steamboat) 26, *26*; Ormen Lange site (unnamed), 72, 73 (see also Ormen Lange main entry); *Pedro Nunes* (formerly *Thermopylae*), 55; Punic (unnamed) site, 89, 93; SS *Republic* (side-wheel steamer) 11, 91, 91–92—ownership of, 91; Roman sites (unnamed), 3, 39, 49, 51, 89; *Sainte-Dorothéa* (Danish merchant ship), 33–34; *Santa Margarita*, (Spanish) 84; *Scourge* (schooner), 27; "Shipwreck D" (amphora carrier), 52, 54; M/V *Stockholm*, 19; *Sud-Caveux* (Roman), 34, 35; HMS *Sussex*, 89–91, *90*, 93, 166; "Tanit," 44; "Target 153," 52; *Thermopylae* (renamed *Pedro Nunes*) 55, *55*; RMS *Titanic*, 19, 36, 101, 146, 156, 162; *U-166* (German submarine), 21; *Vasa* (Russian ship), 55, 164; Vasiliko Bay (Byzantine) site, 46, 47, *47*, 48; *Voronov* (Russian ship), 62, 63, 65; *Vrouw Maria* (Dutch ship), 55; *Wasa* (Swedish navy vessel), 140–41, 149, 150; *Western Empire*, 28–29; Xlendi site, 40, 42–43
Sicily, 35, 36, 40, 43. See also Skerki Bank
Sinop (Turkish seaport), 52
Site Recorder 4 (3H Consulting), 31, 153–54
Skerki Bank (off Sicily), 3, 19, 37
Smøla (island off Norway), 61, 62, 64, 167
Solombalskaya shipyard, 62
sonar: DSL-120 sidescan, 44; Edgetech CHIRP, 91; EG&G sidescan, 25, 57; magenex scanning system, 39; Klein—595, 84—System 2000, 59; Kongsberg scanning, 27, 59, 66, 74, 85, 95; SeaScan PC, 45; SeaMARC IA, 93–94; TOPAS (topographic parametric), 60, 68, 69, 131. See also bathymetry
Sonardyne (company), 86, 90, 117, 150
Spain, 32, 33, 43, 81, 84, 96
Sperre (company), 29, 45, 74
SS *Central America* Discovery Group, 93, 95, 96
SSBL (super-short baseline underwater): accuracy of, 29–30, 127; and ease of operation, 117; technology of, 117–18; and underwater vehicle positioning, 39, 49. See also Kongsberg: SSBL; SBL
St. Petersburg, Russia, 62, 63, 64
stratigraphy, 51, 130, 141, 145, 150
Statfjord (oil/gas field), 167
Statoil (Norwegian state oil company), 61, 62, 65, 167–68. See also Statpipe
Stene, Arne, 62, 63
Stockholm, 55, 57
Stockholm archipelago, 12
Stolt Offshore (company), 69
Straits of Florida, 84
Straits of Otranto, 19
sub-bottom profiling, 68, 70, 76, 77, 102, 108–109, 131; and CHIRP, 130–31—Geochrip II system, 131; and identification of cultural remains, 109, 132; technology of, 108
Sub Sea Oil Services, 35
Submarine Technology (company), 85
submarines: CSS *H. L. Hunley*, 150; *Dakar*, (Israeli) 19, 97; for deepwater archaeology, 18–19; mini-submarines, see submersibles; *NR-1* (U.S. nuclear-powered), 18–19, *18*, 28, 37, 39, 43, 49; USS *Scorpion* (U.S. nuclear-powered), 93; *U-166* (German), 21. See also submersibles (mini-submarines)
submersibles (mini-submarines), 17–19; for excavation tasks, 139; *Aluminaut*, 23; *Cyana* (French), 31, 32, 33; *Griffon* (French) 33; *Johnson Sea Link*, 25, 87; *Nautile* (French), 31, 32, *32*, 33; *Nérée* (French), 34, 35; *PC15*, 36; *Remora* (French), 33, 34; *Remora 2000* (French), 34, *34*, 35, 39; *Thetis*, 51; *Trieste II*, 17. See also submarines
Sweden, 55, 156, 164, 165
Switzerland, 165

Tampa (Florida) 85
Texas A&M University, 28, 29, 35, 98
Thiaki. See Ithaki
Thompson, Tommy, 93, 94
Throckmorton, Peter, 47

Tierra Firme (Spanish fleet), 84
Toisa Vigilant (research vessel), 29
treasure hunting, 7–8, 84, 85, 87, 116, 167; and lawsuits, 95; and looting, 39; problems of, 90
Tripolitania (African coast), 43
Tritech Zip Pump, 70
Trondheim fjord, 57
Trondheim harbor, 57, 58, 59, 60
Trondheim (Norway), 63
Tunisia, 19, 43
Turkey, 52, 165
Turner Broadcasting System, 36
TVSS (television search and salvage system), 24
Tyrrhenian Sea, 23

U.S. Navy, 3, 18, 26, 162; Naval Historical Center, 166, Archaeology Branch of, 166
Ukraine, 54
Ulises (Cuban survey ship), 96
UNESCO, 4; Convention on the Protection of the Underwater Cultural Heritage, 165
United Kingdom, 165: Department of National Heritage, 166; Hydrographic Office, 4; Merchant Shipping Act (1894), 166; and Odyssey Marine Exploration partnership, 89, 166; Protection of Military Remains Act (1986), 166; Protection of Wrecks Act (1973), 166; Royal Commission on the Historical Monuments of England, 166, National Maritime Record in, 166; Royal Navy, 166; and treasure hunting, 166
United States: Abandoned Shipwreck Act, 165, 166; Department of Interior: Minerals Management Services (MMS), 27, 113, 136, 166; National Historic Preservation Act (1966), 166; National Marine Sanctuary Act, 166; National Register of Historic Places, 166; and offshore oil exploration, 166; Sunken Military Craft Act, 166. *See also* NOAA: U.S. Navy
University of Patras, 52
University of Rhode Island, 53
University of Southampton, 130
Uruguay, 165
USNS *Apache* (ocean tug) 25
USS *Housatonic*, 150
USS *Kearsarge* (1860s sloop) 35
USS Monitor Center, 19. *See also* shipwrecks; USS *Monitor*

Val di Compare. *See* Ithaki (Ionian island)
Venezuela, 165
Veolia Environmental Systems, 29
Vera Cruz, Mexico, 96
VETIS (Vehicle Tracking and Information System software), 66
video documentation, 121–23; camera choice for, 121–22. *See also* cameras
Villefranche (bay of), 33
Viosca Knoll (Gulf of Mexico), 28

Walter Hood Shipyards (Aberdeen, Scotland), 55
WebGIS, 39
Westinghouse (company), 17
White Star Line, 19
wood borers, 95; *Teredinidae* (shipworms, teredo worms), 37, 162; *Pholodaceae* (piddocks), 162; *Xyloredo*, 37
Wood's Hole Oceanographic Institute (WHOI), 118; Center for Marine Exploration, 36; Deep Submergence Laboratory, 26, 36; equipment developed by, 21, 27, 36, 76—Argo towed survey system, 19—*Medea/Jason* ROV system, 19, 44—SeaBed (AUV), 51; software designed by, 36, 39
World War I, 19

Xerxes (king of Persia), 51
Xlendi, Bay, Gozo Island, Malta, 40